XX

A DESCRIPTION OF
THE CANALS AND RAILROADS
OF THE UNITED STATES

A DESCRIPTION OF THE CANALS AND RAILROADS OF THE UNITED STATES

COMPREHENDING NOTICES OF ALL THE WORKS OF INTERNAL IMPROVEMENTS THROUGHOUT THE SEVERAL STATES

BY

HENRY S. TANNER

1840

REPRINTS OF ECONOMIC CLASSICS

AUGUSTUS M. KELLEY · PUBLISHERS
NEW YORK 1970

TC
623
.T35
1970

First Published 1840
(New York: T. R. Tanner & J. Disturnell, *124 Broadway*, 1840)

Reprinted 1970 by
AUGUSTUS M. KELLEY · PUBLISHERS
REPRINTS OF ECONOMIC CLASSICS
New York   New York   10001

FROM *A* COPY IN THE COLLECTIONS OF
THE NEW YORK PUBLIC LIBRARY
Astor, Lenox and Tilden Foundations

· · · · · · · · · · ·

S B N   678 00595 8

L C N   68 27678

· · · · · · · · · · ·

Printed in the United States of America
by Sentry Press, New York, N. Y. 10019

A

DESCRIPTION

OF THE

CANALS AND RAIL ROADS

OF THE

UNITED STATES,

COMPREHENDING

NOTICES OF ALL THE WORKS

OF

INTERNAL IMPROVEMENT

THROUGHOUT

THE SEVERAL STATES.

BY H. S. TANNER.

NEW YORK:
T. R. TANNER & J. DISTURNELL,
124 BROADWAY.

1840.

Entered according to Act of Congress, in the year 1840,

BY T. R. TANNER AND J. DISTURNELL,

in the Clerk's Office of the District Court of the Southern District of New York.

A

DESCRIPTION

OF THE

CANALS AND RAIL-ROADS

OF THE

UNITED STATES.

CONTENTS.

	Page.
Advertisement to the first Edition,	9
Introduction,	11
General view,	15
Preliminary remarks on the State of MAINE,	29
Canals, " "	29
Rail-roads, " "	29
Proposed works, " "	30
Canals in the State of NEW HAMPSHIRE,	32
Rail-roads, " "	32
Canals in the State of VERMONT,	33
Proposed works, "	34
Rail-roads in the State of MASSACHUSETTS,	35
Canals, " "	43
Proposed works, " "	43
Rail-roads in the State of RHODE ISLAND,	44
Canals, " "	46
Rail-roads in the State of CONNECTICUT,	47
Proposed Rail-roads, " "	47
Canals, " "	48
Preliminary remarks on the State of NEW YORK,	50
Canals, " "	52
Croton Aqueduct, " "	59
Rail-roads, " "	70
Proposed works, " "	81
Preliminary remarks on the State of NEW JERSEY,	84
Canals, " "	86
Rail-roads, " "	88
Proposed works, " "	94
Preliminary remarks on the State of PENNSYLVANIA,	95
Canals, " "	97
Fairmount Water Works, "	104
Rail-roads, " "	113
Proposed works, " "	132

CONTENTS.

	Page.
Rail-roads in the State of DELAWARE,	147
Canals, " "	148
Proposed works, " "	148
Rail-roads in the State of MARYLAND,	149
Canals, " "	158
Proposed works, " "	159
Preliminary remarks on the State of VIRGINIA,	160
Canals, " "	160
Rail-roads, " "	162
Proposed works, " "	166
Rail-roads in the State of NORTH CAROLINA,	167
Canals, " "	168
Proposed works, " "	168
Rail-roads in the State of SOUTH CAROLINA,	169
Proposed works, " "	171
Canals, " "	171
Rail-roads in the State of GEORGIA,	173
Canals, "	178
Proposed works, "	178
Rail-roads in the Territory of FLORIDA,	179
Canals, "	179
Proposed works, "	179
Rail-roads in the State of ALABAMA,	180
Canals, "	183
Rail-roads in the State of MISSISSIPPI,	184
Proposed works, "	184
Preliminary remarks on the State of LOUISIANA,	185
Rail-roads, " "	188
Canals, " "	188
Proposed works, " "	189
Rail-roads in the State of TENNESSEE,	191
Proposed works, " "	191
Preliminary remarks on the State of KENTUCKY,	192
Rail-roads, " "	193
Canal,* " "	193
Proposed works, " "	193
Preliminary remarks on the State of ILLINOIS,	196
Canals, " "	196
Rail-roads, " "	196
Preliminary remarks on the State of INDIANA,	198
Canals, " "	198
Rail-roads, " "	201
Proposed works, " "	201

* The name of this Canal is erroneously printed "Louisville and *Pottsdam* Canal," it should be Louisville and Portland Canal.

CONTENTS.

	Page.
Preliminary remarks on the State of OHIO,	202
Canals, " "	209
Rail-roads, " "	212
Proposed works, " "	213
ARKANSAS,	214
MISSOURI,	214
Preliminary remarks on the State of MICHIGAN,	215
Rail-roads, " "	215
Canals, " "	219
Proposed works, " "	219
WISCONSIN, Territory of	220
IOWA, Territory of	220
Canals in the Province of CANADA,	221
Rail-roads, "	222
Proposed works, "	222
Condensed Summary of the Canals and Rail-roads of the United States,	223
Glossary of the Scientific, Mechanical, and other terms employed in Engineering,	234
Index,	265

ERRATA.

Page 53 line 24, for 60 or 70 feet, read 70 feet.
Page 53 line 25, for 2 feet, read 3 feet.

ADVERTISEMENT

TO THE FIRST EDITION.

IN consequence of the numerous inquiries relative to the Internal Improvements in the United States, we have been led to draw up, as an accompaniment to the preceding map, the subjoined account of the Canals and Rail-roads which exist in the various states. It will serve, in connexion with the map, to elucidate, fully, the extent, courses, &c., of those great works, to which the attention of strangers, in common with our own citizens, is directed. In the arrangement of the several topics, particular care has been taken to embody, under the head of each state, *all* the canals and rail-roads which exist in it; together with such facts as were deemed generally useful. Among these will be found statements relative to the designation; points of commencement and termination; general course; locality; length; point of greatest elevation; ascent and descent; capacity (width, depth, &c.); number of locks or inclined planes; of dams; of aqueducts or viaducts; of tunnels; of bridges; of sluices; of culverts; cost of construction; present condition; proprietors, &c., of each canal and rail-road in the United States, so far as the requisite data have been obtained.

These items being considered most important in reference to the subject generally, all other matters, especially those which relate to legislative proceedings, and other preliminary operations, found in most other works of this sort, have been purposely omitted as comparatively uninteresting. Our chief aim has been to condense into a small space as much practical information on the various subjects brought into view, as a due regard to satisfactory results would allow. Should errors be detected by those who are familiar with the details of any of the works here mentioned, (and that errors may be found we must admit,) we beg leave to request the communication of such facts as will serve to correct in a future edition of this work, whatever may be found defective in this.

INTRODUCTION.

To AN American sincerely interested in the welfare and prosperity of his own country, it must be peculiarly gratifying to reflect on the rapid advancement in the great work of internal improvement, by canals and rail-roads, which, during the present age, has been made in the United States. Whatever may be the condition of other portions of the civilized world with regard to these important modes of intercommunication, we, in this country, may boast of our superiority, not only in the extent to which the system has been carried here, but also in the admirable formation of its various lines; and, what is of infinite importance, in the structure and management of the locomotive engine. Among us, the spirit of improvement is no longer confined within the limits of our populous states; but, like their restless inhabitants, has found its way to the remotest corners of the republic, where its influence is equally felt and appreciated, and where the system will become as universal as it is beneficial.

This general extension, actual and prospective, of canals and railways in the United States, is one effect of that enlightened constitution of government, which the Revolution has confirmed to us, and which impresses an indelible mark of distinction between the enterprizing American and the plodding inhabitant of the other hemisphere. In most other countries, the great mass of the people, being destitute of wealth, have but little influence, and still less power to effect important objects; hence every work requiring large expenditure, must be accomplished by the wealthy few, whom it is well known, do not always consist of the most enterprising portion of a community. Here it is essentially different; wealth and information being more generally diffused among the great body of the people, they possess and exercise a powerful influence in all affairs of a public nature, and of course, claim a large share of attention. To the *people* of the United States then, is the great system of internal improvement confided.—To this system, therefore, men of learning and influence now direct their attention, as the *people*, in the most comprehensive sense of the term, are to derive the advantages, which must result from its general adoption.

Books divested of all superfluous technicalities, on the construction of canals and railways, reduced from that size which suited only the purses of the rich, and adapted to the wants of the practical engineer, are greatly needed. It is to works of this description, not only on the subject of internal improvement, but all others, more than to those of an expensive class, that our countrymen owe that superior improvement which distinguishes them from the people of all other countries.

To promote and advance the knowledge of this system, is the leading object of the present undertaking. No subject at this time appears more important than that which we have chosen, and, if treated in a way that may render it generally useful, cannot fail to elicit the approbation of its friends.

The knowledge of the existing canals and railways of our country, though not absolutely necessary for all classes of the community, it must be allowed, is that which every well educated person is now expected to possess; and hence, books which describe works, such as we have mentioned, promise the best means for obtaining this knowledge.

The description of those works now offered to the public, differs in several points from other books on the same subject. Besides exhibiting an easy, distinct, and systematic account of the existing works of internal improvement, we have endeavoured to describe those merely contemplated, which are likely to be constructed soon. The value and importance of canals and railways depend on a combination of circumstances, which reciprocally affect each other. There is a nearer connexion between works situated apart and distant from each other, than most people seem to apprehend. In a work of this kind, the object of which is to develop the moral, political and commercial effects of the system of internal improvement in our country, none of these topics should remain unnoticed. The omission of any one of them would, in reality, deprive us of a branch of knowledge, not only interesting in itself, but which is absolutely necessary to enable us to form a just conception of the subject in general. We have therefore thought it necessary that our work should embrace the history and present condition of the canals and rail-ways in every state and territory of the union, with the particulars and details belonging to each. This will, on reflection, appear necessary, when we consider the powerful influence of the system upon the habits and pursuits of a vast number of our people.

These subjects, indeed, till of late, seldom found a place in geographical works; and, even where they have been introduced, are by no means treated in such a manner as to afford the desired information. Neither is this to be altogether imputed to inattention on the part of geographical writers. The subject is too important and extensive to admit of satisfactory description in a work not exclusively devoted to it. Such is the difficulty of obtaining the requisite data for a just representation of the whole system, that no one, who relies exclusively upon the ordinary means of acquiring them, can do full justice to the subject.

The general geographer, then, who could only employ the materials placed in his hands, could scarcely be expected to give that comprehensive view of it, which its increasing importance demands. In the progress of improvement, however, the means of acquiring correct information have greatly increased, and nothing is now wanting but a moderate share of diligence and patient research, to enable any competent writer to describe satisfactorily every important work in the country. Without manifest impropriety, we could not but avail ourselves of those additional facilities, by the aid of which, we have been enabled to give a more copious and perfect account of what is comprehended in the terms "*Internal Improvements*" than has hitherto appeared.

In considering the present and future condition of the several states, few circumstances are more important than their mutual intercourse. This is chiefly promoted by commerce, the all powerful agent in the economy of modern states; and with a constant reference to this consideration, have we prepared the present account.

Having, through the whole work, mentioned the principal towns affected by any canal or railway, we have thought it necessary, for the satisfaction of our readers, to insert a map of the United States, including the various lines, coloured in such a manner as to distinguish between the canals and railways. By this arrangement we afford the opportunity to the reader, of comparing, not only the length and location of the several works as they now exist, but also those merely contemplated. The map, which exhibits a general view of the country, will illustrate satisfactorily the descriptive parts.

In the execution of our design, we have uniformly endeavoured to observe order and perspicuity. Brevity and truth rather than refinement of style, have been our aim: whether we have succeeded in our purpose, is not for us to determine.

What has enabled us to embody so many facts, within the narrow bounds of this work, is the omission of all extraneous matter, such as the legislative proceedings and other acts which are necessary in the incipient stages of a work, but are perfectly useless so far as any practical benefit is concerned.

In describing the various canals and rail-roads, we have been more or less diffuse, according to their importance, with regard not only to their magnitude, but also to their usefulness as connecting links in a general series.

In treating of such a variety of subjects, some less obvious particulars have, doubtless, escaped our notice. But if our plan be good in the main, and the prominent outlines correctly delineated, the candour of the reader, we trust, will excuse imperfections which are unavoidable in a work of this description.

We cannot, without exceeding the prescribed bounds, say much respecting the other parts of our work. The maps and diagrams, which were originally drawn and since revised with great care, will, it is believed, give satisfaction to those who consult them. With regard to the subject generally, we have on all occasion of doubt or uncertainty, resorted to the most approved sources for information, to which access could be had.

INTERNAL IMPROVEMENTS

OF THE

UNITED STATES.

GENERAL VIEW.

In order to comprehend truly, the full extent to which the system of internal improvements has been carried in the United States, and correctly to estimate its future progress, it will be useful to take a general view of the origin and present condition of those leading works, which from their continuity or connection with others, partake more of a national, than of a merely local character.

In this class may be ranked all such works as do at present, or may hereafter unite, and thus form a continuous line of communication between distant portions of the country; such, for example, as the Camden and Amboy; Philadelphia, Wilmington and Baltimore; Baltimore and Washington, and other rail-roads. In this point of view a canal or rail-road, however insignificant in regard to length or other characteristic, assumes an importance far beyond its intrinsic value, when considered only in reference to its individual capacity.

The history of internal improvements in the United States for the last twenty-five years is, indeed, replete with interest. Within that eventful period, we have witnessed the commencement, progress and ultimate completion of those great works, the effects of which are every where developing themselves in the rapid increase and extension of canals and rail-roads throughout the civilized world. The most sanguine anticipations of the promoters of our system of internal improvements, have fallen short of the actual results which their firm adherence to that system has produced. Judging from the past, and present manifestation of public spirit, every where exhibited, we may anticipate a still more rapid and effectual progress in the interior improvements of the United States, than at any former period. Explorations, surveys and other measures preparatory to the construction of rail-roads, more particularly, now occupy the attention of legislative bodies, as well as companies and individuals throughout the United States. This country indeed appears to have led the way in the most useful undertakings of this description; and the economy with which most of them have been executed, when compared with the cost of similar works abroad, is matter of surprise to all, and deserves the attention of those who are practically engaged in the work of internal improvement.

The improvement of the channels of intercourse between the various sections of the same country, has always been considered one of the first duties of the government; and we find that roads and other means of intercommunication, are more or less improved in all civilized countries; but it is in commercial and manufacturing communities, where roads and canals receive their due share of national attention. Rail-roads and canals are, in the most emphatic sense, labour-saving machines, and it must be admitted, that their construction and improvement in the United States, have progressed with a celerity and magnitude, greatly exceeding the increase of population or the extension of settlements. In the following list of those works, we have endeavoured to give an account of all the canals and rail-roads, either completed or in progress in the United States. It will serve to exhibit the vast and increasing interests already created by the introduction and extension of internal improvements throughout the country; and satisfy, in some degree at least, that desire for detailed information on the subject, which is now manifested by every intelligent reader. It, moreover, shows that American energy and perseverance can effect the most important and invaluable achievements. In this age when the accomplishment of one great design is taken as an incentive to undertake, and as a demonstration of the practicability of executing other plans, still more magnificient, we may anticipate the most splendid results; and in less than a quarter of a century, locomotives and their attendant trains will be found traversing regions far beyond the present haunts of civilized man. The pre-eminent advantages of canals and rail-roads have been established by the unerring test of experience, and the entire country can now bear testimony to the superior facilities which they afford, in promoting useful intercourse. As to the purposes of beneficial communication, they diminish the distance between places, and thus encourage the cultivation of the most extensive and remote parts of the country. They create new sources of internal trade and augment the advantages of natural channels. Internal navigation and the facilities of intercommunication which rail-roads create, may in fact be regarded in the same light as exterior navigation, when viewed in reference to the great family of mankind. As the oceans connect the nations of the earth by the ties of commerce, and the benefits of communication, so do canals and rail-roads operate upon the inhabitants of the same country. The arguments in favour of internal improvements apply with peculiar force to the United States.

The immense effects which a connected system of improvements between the East and West, must produce in securing the trade of the latter to the Atlantic states, and in cementing that bond of union upon which the safety of our institutions depends, would furnish, upon the proper occasion, a subject of interesting and profitable inquiry. But this is not the appropriate time to expatiate upon the brilliant prospect which would open upon us were such a system perfected and brought into active operation; nor to attempt to measure the consequences which its completion would produce in augmenting the stream of inland commerce, which would flow from its introduction.

Most of the rail-roads, which have hitherto been constructed in the United States, were designed merely to connect certain local points without reference to any general system. Notwithstanding this want of forethought on the part of their projectors, many of those roads now form parts of extensive thoroughfares, and others will be ere long made available in completing other leading avenues. The fact that the respective sections are under the management of companies wholly independent of each other, and subject to different, and in some cases, conflicting regulations, may perhaps lead to unpleasant results. Some of the companies fore-seeing this, have united their interests and by this means avoided the effects of a want of unanimity in their operations. Thus in the route between Philadelphia and Baltimore there were no less than four chartered companies: these now form a single interest under the management of a board of directors chosen by the stockholders of each in joint ballot.

The first great chain of rail-road, of which those just mentioned form a part, is that commencing at Portsmouth, in New Hampshire, and extending with an occasional interval through the Atlantic states, to Pensacola in Florida. From Portsmouth the Eastern Rail-road extends to Boston, whence the line is continued, by the Boston and Providence Railroad to Providence, where it meets the rail-road to Stonington, in Connecticut. From Stonington, after crossing Long Island Sound to Greenport, on Long Island, the line is resumed, and proceeds to Brooklyn, opposite New York, by the Brooklyn and Long Island Rail-road, about 28 miles of which are completed and in use, the remaining 72 miles are now in progress. Crossing the East river to New York, and thence over the Hudson to Jersey city, the line is continued by the New Jersey Rail-road to New Brunswick, thence by the Trenton and New Brunswick Rail-road to Trenton, and thence to Philadelphia by the Philadelphia and Trenton Rail-road. From Philadelphia it proceeds to Baltimore by the Philadelphia, Wilmington and Baltimore Rail-road, and thence to Washington by the Washington branch of the Baltimore and Ohio Railroad. The road from Washington to Fredericksburg in Virginia, though proposed, is not yet commenced. At Fredericksburg the line is resumed and proceeds to Richmond by the Fredericksburg and Richmond Rail-road, thence to Petersburg by the Richmond and Petersburg Rail-road, thence by the Petersburg and Roanoke Rail-road to Gaston in North Carolina, thence by the Raleigh and Gaston Rail-road to Raleigh, whence it is proposed to construct a rail-road to Columbia, in South Carolina. From Columbia, by the Columbia branch of the South Carolina Rail-road, the line is conducted to Branchville, and thence by the main line of the South Carolina Rail-road to Augusta in Georgia. At Augusta commences the Georgia Rail-road, which extends to DeKalb county, whence a road to West Point, on the Chottahooche, is in progress.

GENERAL VIEW. 19

From West Point the line proceeds along the Montgomery and West Point Rail-road to Montgomery in Alabama, and thence by the Alabama, Florida and Georgia Rail-road to Pensacola in Florida. In the entire length of this extensive line, there are but four sections wanted to render it complete, viz. one from Greenport to Hickstown, Long Island; one from Washington to Fredericksburg; one from Raleigh to Columbia, and one from De Kalb in Georgia to West Point. The aggregate length of these rail-roads, nearly all of which are completed, and in use, is 1600 miles. Should the state of Virginia execute her projected rail-road from Richmond, via Abingdon, to the Tennessee line, a route to New Orleans will be effected by means of the Highwassee, Knoxville and Nashville, and the New Orleans and Nashville Rail-roads, now in progress.

The second great thoroughfare commences at Boston, and proceeds to Worcester by the Boston and Worcester Rail-road; at Worcester the line is continued by the Great Western Rail-road to West Stockbridge, and thence to Albany by the Albany and West Stockbridge Rail-road, (now in progress.) From Albany the line proceeds along the Hudson and Mohawk Rail-road to Schenectady; thence by the Schenectady and Utica Rail-road to Utica; thence by the Syracuse and Utica Rail-road to Syracuse; thence by the Syracuse and Auburn Rail-road to Auburn; thence by the Auburn and Rochester Rail-road (in progress,) to Rochester; thence by the Tonawanda Rail-road to Attica; and thence by the Attica and Buffalo Rail-road to Buffalo. This line is also nearly completed. It is 530 miles in length, and is one of the most important highways in the United States, especially that portion of it which extends from Albany, westward.

The third route is that by the New York and Erie Rail-road, about 450 miles in length, only a part of which is in course of execution, but from the measures that have been adopted, its early completion may be expected.

The fourth route commences at Philadelphia, with the Reading Rail-road to Port Clinton; thence by the Danville and Pottsville Rail-road to Sunbury, or the Little Schuylkill and Susquehanna Rail-road to Williamsport; and thence by the Sunbury and Erie Rail-road to the town of Erie, on Lake Erie, 420 miles.

The fifth route consists of rail-roads and canals. It commences at Philadelphia, and proceeds to Columbia, on the Susquehanna, by the Philadelphia and Columbia Rail-road. The central division of the Pennsylvania canal takes up the line at Columbia, and proceeds to Hollidaysburg, at the eastern base of the Allegheny mountain, which is crossed by another rail-road, (the Allegheny portage,) extending to Johnstown on the western declivity of the same ridge. At Johnstown commences the western division of the Pennsylvania canal, which terminates the line at Pittsburg. Length, 394 miles.

The sixth route extends from Baltimore to Wheeling, on the Ohio river, by the Baltimore and Ohio Rail-road, 280 miles in length.

The seventh route commences at Richmond, in Virginia, by the James River Canal, which extends to Covington, in Allegheny county; thence by Rail-road to Loup shoals in the Kanawha river; and thence by the Kanawha, the navigation of which has been improved by dams and locks, to the Ohio river.

The eighth route, 718 miles in length, extends from Charleston by the South Carolina Rail-road to Columbia, and thence by the proposed Louisville, Cincinnati and Charleston Rail-road, to Cincinnati, Ohio; and the

Ninth route extends from Savannah, in Georgia, via the Central Rail-road, to Macon, and thence by the Alabama roads to Pensacola.

All these routes it will be perceived, are connected with the great Atlantic line, which may be regarded as the main artery in the system. The minor canals and rail-roads, in turn, are mere ramifications of those branches, and as such, augment the revenues of the whole.

In addition to these extensive avenues of trade on the Atlantic border of the United States, the western states are busily employed in constructing lines of rail-road, not less than two thousand miles in extent, in order to unite the navigable streams with the great lakes. The cost of these works, most of which are in actual progress, will exceed $50,000,000.

The circumstance, moreover, which is particularly important, is that the public works in each of those states, are arranged on a harmonious plan, each having a main line, supported and enriched by lateral and tributary branches, thereby bringing the industry of their whole people into prompt and vigorous action, while the systems themselves are again united on a more extensive scale, in a series of systems comprising an aggregate length of more than 2,000 miles. The several sections of this extensive work are now in a train of rapid construction. Ohio, Indiana, Illinois and Michigan, are straining every nerve to perfect their various lines; so that it may be confidently predicted, that within seven years from this time, nearly the whole inland trade of that wide spread region will be conducted through their canals and rail-roads into those of the Atlantic states, and thus avoid the circuitous course of the Mississippi, and the more dangerous navigation of the Florida Gulf.

In the south, the people, alive to the importance of the subject, are making great exertions to complete the lines of rail-road already commenced; and in a few years, Virginia, the Carolinas, Georgia and Alabama, will be intersected by rail-roads in nearly every direction.

By the completion of the James River Canal and Rail-roads, the Louisville, Cincinnati and Charleston Rail-road, the Georgia and Western and Atlantic Rail-road, the towns on the Atlantic will be so united with those of the interior, as to secure to the latter not only a continuance of the trade which they now enjoy, but will contribute to augment in no slight degree, the future commerce of those states.

Perceiving that the completion of the northern system of rail-roads would deprive the southern seaports of their accustomed trade, unless its effects were counteracted by the construction of similar works in the south, her citizens lost no time in commencing a corresponding system, by which the threatened contingency might be averted, and her comparative standing maintained. This system is now rapidly advancing towards completion, many of its lines being at this moment in successful operation, while others are on the point of being opened for the public.

Thus, it will be perceived, that the entire surface of our country is now, or will be shortly, intersected by canals and rail-roads in almost every direction; and the west will ere long present a spectacle equally cheering to the friends of internal improvements. As facilities of intercourse, the moral effects of the general introduction of rail-roads and canals can never

GENERAL VIEW.

be duly appreciated. Considered as means of revenue, merely, it is doubtful whether they can be made to yield an interest equal to that derived from most other investments. With regard to the canals of any country, taken in the aggregate, their average income falls considerably short of the current interest of the country. Some of the canals of England, those of Coventry, Erwash and Laughboro, for example, yielded in 1822, an average annual interest of upwards of one hundred and twelve per cent. on their cost; whilst others, scarcely defray their ordinary expenses. The average receipts from the New York State Canals for the last three years, have yielded an interest on these of about eight per cent. And the principal canals of Pennsylvania, for the same period, have produced nearly six per cent. The tolls for the last fiscal year, ending on the 31st October, 1839, were on all the canals, $542,886 63; on rail-roads, (Columbia and Portage,) $319,622 88; on motive power, $280,123 53; total, $1,142,633 04, which exceeds the annual aggregate of the preceding year, by $151,380 62. The rail-roads throughout the country, will, no doubt, prove hereafter to be more productive, than the canals; though, according to a statement drawn up by Mr. De Gerstner, the interest on the capital invested in rail-roads in the United States in 1839, does not exceed five and a half per cent. per annum.

This result is based upon the conclusion drawn from personal investigation, that the aggregate cost of our roads, is $20,000 per mile, including buildings and all requisite aparatus. The same gentleman states, that the average amount paid by each passenger conveyed on the American rail-roads, *is five cents per mile;* that passengers are conveyed with a speed of from twelve to fifteen miles per hour, stoppages included; that there are on an average 35,000 *through* passengers, and 15,000 tons of merchandize, carried annually over the American railroads; that the expenses per mile of travel are one dollar; that the average number of passage trips per year is 875; that the expenses per passenger per mile are two and a half cents; and that the annual current expenses for working the American rail-roads, are $1,950, or $63\frac{61}{100}$ per cent. per mile, on the gross income. The foregoing estimates are founded upon the condition of our rail-road establishment as it existed in the year

GENERAL VIEW. 23

1839, when many of the roads which were taken into the account, had scarcely commenced operations; whilst others were in such a state as to prevent their connection with established lines, upon which income materially depended. The average cost per mile of rail-road executed hereafter, will, undoubtedly, be greatly reduced, especially in the south, where a large proportion of the new rail-roads are now in progress. With a diminished cost of construction, and an augmented revenue, the result of an union of the present disjointed parts of the system, and increased experience on the part of our engineers, an average revenue, equal to seven per cent. per annum on the aggregate cost of our rail-roads, may be confidently anticipated for several years to come.

The gross income of the Pennsylvania state rail-roads for the last year, was, as we have seen, $319,622 88; from which deduct expenses, $27,941 34, and we have $291,681 54, as the net amount of revenue from this source, or about five per cent. on the total cost of these works; which, owing to their peculiar construction, were attended with unusual expense.

With regard to the abstract question of revenue it is obvious that a large portion of the immense sums invested in canals and rail-roads in the United States, will fail in producing the anticipated results. Visionary enterprises of all sorts, are the distinguishing characteristics of the times; and the almost infinite variety of schemes, which of late have been pressed upon public attention, and adopted without due caution, have in some instances, resulted in the diversion of funds from objects of undoubted utility and advantage, to those of an opposite character:—whilst the rate of interest in this country continued as it has been for many years past, none but the most promising enterprises should have been undertaken by individual companies. The mode of improvement and its fitness to the purposes for which it is designed, are considerations to which little regard has been paid in deciding upon the location of some of the public works in the United States. Hence the numerous failures and the consequent withdrawal of public confidence in such investments generally.

"*Lines* of communication," we quote from a spirited article in the New York Review, " may be judicious, and the plan of the improvement be otherwise; or, if the line and plan be both eligible, an error may be committed in making the improvement incommensurate, or more than commensurate, with its objects. We need only refer, in illustration of our position, to the opposite plans of connection by canal and rail-road, or to the case of a single or double track rail-road. It will be easily perceived, that a trade or travel which would pay a large profit on the stock of a rail-road, might not be commanded by a canal, and *vice versa;* and that whilst it would be a great error to make a double track rail-road where a single track rail-road would suffice, it would not be a less error to construct a road graded for a single track only, where a double track would be required. In cases where either a canal or rail-road may be adopted, it is obvious that a great error would be committed by selecting that least suitable to the circumstances of the case, and the trade to be accommodated. In our opinion, and in the opinion of engineers who have given most attention to the subject, there are three cases in which rail-roads present decided advantages over canals:—

1. *Where persons or articles of great value are to be transported.* In this case, the saving in time becomes a matter of much more moment than any increased cost of rail-road over canal transportation.

2. *Where great difference of level are to be surmounted.* In this case, the delay and risk of interruption from a large amount of lockage, is far beyond what would be occasioned by inclined planes overcoming the same elevation on a rail-road, whilst the same causes increase the cost of canal transportation to so nearly an equality with that of rail-roads, as to entitle the latter to the preference.

3. *When, no matter how bulky or ponderous the tonnage in proportion to its value, the trade is principally a descending one, and a profile can be had for the rail-road giving equal, or nearly equal, facilities to the power employed in both directions.* In this last case, rail-road transportation, more advantageous in other respects, becomes even cheaper than that on canals, in consequence of the fall in the line, which in the case of the canal would present a serious impediment, by the lockage it would occasion, becoming an auxiliary to the power employed on the rail-road. There are but few cases, of course, where *the most desirable* profile has been had for a rail-road, the descent being

in most cases, either somewhat more or less than would be in theory the most advantageous. One of the most remarkable approaches to it seems to have been made in the case of the rail-road now under construction between Philadelphia and the anthracite coal districts, at the head of the Schuylkill, in Pennsylvania. On this work the graduation is said to vary between a level and a descent of nineteen feet per mile, in favour of the downward trade. It is thought the load of ordinary locomotive engines on it will average from one hundred and fifty to two hundred tons net, or three hundred tons gross, travelling at the rate of from ten to twelve miles per hour, and that the whole expense to the company, of transporting coal on it, will be only fifty-three cents per ton the whole length of the railroad (ninety-six miles,) or little more than half the ordinary rate of freight on canals of the same length."

The above propositions cannot of course, be laid down absolutely, or without reference to the cost of the one or the other improvement: but, in the case first cited, it would require a very great difference of expense to outweigh the decided superiority of the rail-road in point of expedition; and in the second and third cases, the item of lockage will generally enhance greatly the cost of a canal beyond that of a rail-road, and turn the scale more decidedly in favour of the latter improvement.

" Supposing the appropriate description of improvement to be adopted, it may be executed on a plan incommensurate, or more than commensurate with its object. In the way of canals, the Union Canal in Pennsylvania and the Chesapeake and Ohio Canal, are examples of the opposite errors on this subject; it being now generally conceded, that the former, had it been constructed of larger section, and with larger locks, so as to have admitted of the passage of such boats as are used on the Schuylkill and Susquehanna Canals, could not have failed to have been a productive work; whilst the dimensions of the Chesapeake and Ohio Canal seem not to have been warranted by any trade to be anticipated on it.

" To make a double track rail-road when a single track railroad is sufficient, is evidently an equally grave error; but not more so than others which may arise in works of this description, from inattention to their objects and the best means of attaining them. A few words will be sufficient to show what serious mistakes may be committed on this head. It has been observed that short ascending and descending grades present no serious impediment, *at high velocities*, to the most valuable application of locomotive power. They are, consequently, not seriously objectionable on rail-roads for passenger transportation exclusively, on which high velocities will always be aimed at, and where the whole adhesion of the engine is not required; it is otherwise, however, on rail-roads designed for freight. On these it is most essential that the company should be able to transport *cheaply*, and this, in rail-road transportation, is only to be effected by carrying *heavy loads at slow velocities*. Now, at very reduced rates of speed, the limit to the useful effect of the engine is its adhesion, and the load which this will admit of its taking becomes diminished, therefore, precisely as the maximum grade of the road increases, whether this be a longer or shorter one. This being the case, it follows, that on roads destined for the transportation of freights, a much greater expense is justifiable to avoid undulations or to diminish the grades of the roads, when ascents and descents are unavoidable, than on roads destined for passengers. On these last, undulations not exceeding twelve or fifteen feet per mile would be scarcely an objection; whilst on roads designed for the transportation of freight at low velocities, they would almost diminish the load conveyed one-half. In these last, however, the expense which is justifiable to improve the grades of the road, must be materially influenced by the amount of trade to be anticipated on it. If this is but limited, it is, of course, better that the road should be less perfect, and cost of transportation in consequence enhanced, than that an increased first cost, more than commensurate with the object, should be incurred in reducing it; and, on the other hand, the largest outlay may properly be incurred in improving the profiles of a rail-road, in a case where the trade to be accommodated is of proportionate magnitude."

From the above reasoning, which is, we think, quite conclusive, it is manifest that engineers should exercise the utmost caution, lest by any inadvertence, they should expose their employers to irretrievable losses and themselves to the mortification which must result from the misapplication of their professional resources. Whether the internal improvements of the

United States will prove sufficiently productive to ensure the ultimate liquidation of the debt incurred in their construction, is a question of secondary importance, when viewed in connection with the moral, political and commercial advantages which must flow from their use. The advocates of the system were, doubtless, influenced by higher views, than those of a merely pecuniary nature, and the fruits of that system are now developing themselves in the rapid extension of our settlements; in the increased value of agricultural products; in the improvement of the social condition of the people, and in promoting that friendly intercourse between them which is the surest guaranty for the preservation of our institutions.

Merely counting the cost and estimating the probable revenue, without regard to the advantages of internal improvement, in a moral point of view, are inconsistent with those enlightened views, which should enter largely into such calculations. We shall therefore abstain from any further attempt, satisfied that a mere statement of *facts* connected with the system is all that is necessary to enable the reader to draw conclusions to which those facts are so well calculated to lead.

We have prepared a brief explanation of the terms used in describing the elementary parts of rail-roads and canals. It will be found in the end of the volume. This explanation may perhaps be deemed by some, as a work of supererogation: but it must be borne in mind by such persons, that all readers are not engineers : and to the uninitated, we feel persuaded, that an explanation, divested of all technicality, will prove not only useful, but indispensable to a clear understanding of the text. We embrace this occasion also to state that we have given, in its proper place, an elaborate account of the Columbia and Philadelphia Rail-road with its appliances, under an impression that it afforded a complete insight into the construction, operation and uses of rail-roads in general. The variety of material and its diversified combination; the various forms and principles of construction; the numerous kinds of rails employed in its several sections, and the modes by which they are secured, are perhaps better developed on this road than any other we could have selected. Our precision in this instance may, possibly, fatigue some readers, but in describing a road whose formation and materials partake so much of the nature of those of most other rail-roads, we determined to enter thoroughly into all its minutia, and thus avoid the unprofitable task of describing in detail, other roads similarly constructed, of which the Columbia Rail-road affords a satisfactory model. By the adoption of this course, the frequent repetition of matters already described, is dispensed with. In treating of other roads, particular details have been scrupulously avoided, except in cases where they differ essentially in construction from the one we have selected for description: among these are the Great Western Rail-road of Massachusetts, and some of the Southern Rail-roads, which, from the nature of the country traversed by them, required a mode of structure almost unknown in the north.

INTERNAL IMPROVEMENTS.

MAINE.

In 1836 the legislature of this state directed the organization of a "Board of Internal Improvement." It consists of the governor, who is *ex-officio*, president of the board; members of the council; and the land agent. This board is charged with the execution of all works, in any manner connected with the internal improvement of the state; and especially, such canals and rail-roads as the legislature may authorize. With two or three exceptions, little has yet been done in Maine, towards the extension of her improvement system. The Bangor and Orono Rail-road, and the Cumberland and Oxford Canal, are the only works of this description, completed in this state.

CANALS.

CUMBERLAND AND OXFORD CANAL, extends from tide water near Portland, to Sebago Pond, a distance of $20\frac{1}{2}$ miles. By a lock in Songo river, the navigation is prolonged into Brandy and Long Ponds, a farther distance of 30 miles. The canal portion of this work, is wholly in Cumberland county. Length $50\frac{1}{2}$ miles; 34 feet wide at the surface, 18 at the bottom, 4 feet deep; course, north-west; 26 wooden locks; completed in 1829; cost $250,000; cost per mile $12,500; constructed by a joint stock company which possesses banking privileges.

RAIL-ROADS.

BANGOR AND ORONO RAIL-ROAD. Ten miles in length; commences at Bangor, and passes along the right bank of the Penobscot, to Orono, both in Penobscot county. The company was incorporated in 1835, and the road opened for public use in 1836.

CALAIS AND MILTOWN RAIL-ROAD, in Washington county, about 5 miles in length.

PORTLAND, SACO AND PORTSMOUTH RAIL-ROAD. A company for the construction of this road, which is designed as an extention of the Eastern Rail-road, now nearly completed to Portsmouth, was incorporated in 1837. This line commences on the Salmon Falls river, opposite Portsmouth, in N. Hampshire, and thence proceeds in a general N. N. E. direction, through the towns of Wells, Kenebunk Port, and Saco, in York county, and terminates at Portland, in Cumberland. Length 48 miles; estimated cost $781,507 72, or $216,281 61 per mile, exclusive of land damages and fencing.

PORTLAND AND BANGOR RAIL-ROAD. The route proposed is through the counties of Cumberland, Lincoln, Waldo, and Penobscot, 132 miles in length; and cost, as estimated by the examining engineer, $2,475,000.

PORTLAND AND QUEBEC RAIL-ROAD. In the month of July, 1835, an officer of the United States Topographical corps, commenced a reconnoissance of the country between the sea-coast of Maine and the river St. Lawrence, with a view to the construction of a rail-road "from Portland or some other point on the sea-board of Maine, to some point on the borders of Lower Canada, in the direction of Quebec." He completed his examination to the entire satisfaction of the authorities of Maine, and extended his explorations beyond the boundary line, so as to exhibit a connected view of the whole ground, from the coast of Maine to the city of Quebec. The prolongation of the survey beyond the limits of the state, had been assigned by the Canadian Government, to an officer of the royal engineers.

Several routes were examined, all of which were deemed practicable. The first commences at Portland, proceeds through the towns of Falmouth, Gray, Poland, Paris, Rumford, Andover, and some others, and descends into the vallies of Arnold's River and Lake Megantic, Chaudiere and Echemin rivers, to the right bank of the St. Lawrence, opposite Quebec. Length 277; and estimated cost $6,349,671.

The second route is from Wiscasset, along the valley of Sheepscut river, by Webber's Pond, thence over to that of the Sebasticook, enters the valley of the Kenebec, which is pursued to Currituck falls. Ascending the valley to the mouth of Dead River, it passes Wilson's and Cold Streams, Parlin Pond, &c., to Moose river, and thence to the boundary near the Monument; from this point it descends rapidly to the De Loup valley, and along this valley to that of the Chaudiere, where it unites with, and pursues the course of the first line, to its termination on the St. Lawrence. Length 246 miles; and estimated cost $5,419,626.

The third route commences at Belfast, proceeds through the valleys of Wescott March, Halfmoon, and Sandy streams, and thence to the Sebasticook, by Sibley Pond, to the Kenebec, where the line intersects the first and second routes, and becomes identical with them. Length 227 miles; and estimated cost of construction $4,906,151. Average cost per mile; first route, $22,923:—second route, $22,030:—and third route, $21,613. Route No. 3, from Belfast to Quebec is recommended as the most eligible as well as the most economical of the three.

PORTLAND AND DOVER RAIL-ROAD. About 46 miles in length, is proposed. It will pass through the towns of Buxton, Alfred, and South Berwick, where it crosses into New Hampshire, and thence proceeds to Dover, about five miles from Berwick.

PORTLAND AND AUGUSTA RAIL-ROAD, about 60 miles in length, is proposed.

SEBASTICOCK AND MOOSE HEAD LAKE CANAL, 100 miles long, is also proposed.

Aggregate length of canals completed in Maine, 50.50 miles.
" rail-roads, " 10.00 "

NEW HAMPSHIRE.

RAIL-ROADS.

EASTERN RAIL-ROAD. The extension of the Massachusetts Eastern Rail-road, into New Hampshire, commences on the state line, about six miles from Newburyport, and crossing Salisbury marsh, proceeds northward, leaving Hampton Falls to the left, and passing through Old Hampton Village, enters Portsmouth and there unites with the Portland, Saco and Portsmouth Rail-road in Maine. Length of the New Hampshire section, 15.47 miles.

This section of the road is constructed upon principles similar to those of the Massachusetts line. It consists of four straight lines, connected by curves of 5280 feet radius. Maximum grade 35 feet per mile. To be finished in June, 1840.

NASHUA AND LOWELL RAIL-ROAD, extends from Nashua Village in Hillsboro county, N. H., to Lowell on the Merrimac, in Massachusetts, and is a prolongation of the Boston and Lowell Rail-road. It was opened for use on the 25th of Oct. 1838. On Jan. 23, 1839, the Company had expended upon the work $285,052 26. Two companies, one in New Hampshire and the other in Massachusetts, were incorporated for the purpose of constructing the Nashua and Lowell Rail-road: these were subsequently united under the title of the Nashua and Lowell Rail-road Corporation. Length of the road about 15 miles. Its extension to Concord is proposed and a company organized for the purpose.

CANALS.

Bow CANAL, along Bow Falls, three-fourths of a mile in length, four locks; twenty-five feet fall; finished in 1812; cost $25,000.

VERMONT. 33

Hookset Canal, at the Hookset Falls of Merrimac ; 825 feet long : three locks : sixteen feet fall : cost $17,000.

Amoskeag Canal, at the falls of Amoskeag, in the Merrimac, seventeen hundred and sixty yards in length : nine locks : forty-five feet fall : cost $50,000.

Union Canal, passes seven falls in the Merrimac : length including slack water navigation, nine miles : seven locks : cost $50,000.

Sewall's Falls Canal, completed in 1837 : length one-fourth of a mile.

Aggregate length of canals in New Hampshire, 11.13 miles.
" " rail-roads " 30.47 miles.

VERMONT.

With the exception of some small canals, designed to overcome obstructions in the navigation of the Connecticut, there is no canal worthy of notice in this state. These improvements consist, principally, of dams with locks from one pool to another. The following are the chief.

CANALS.

White River Canal, Waterquechy Canal, and Bellows Falls Canal, the latter of which is about 880 yards in length, with nine locks ; and a fall of 50 feet. It is sufficiently capacious to admit the passage of all such vessels as navigate this part of the river, which is thus rendered navigable for about one hundred and twenty miles above the lower falls.

The legislature, during the session of 1835, incorporated companies for the following rail-roads ; nothing, however, has yet been done towards their execution.

RAIL-ROADS.

The Connecticut and Passumpsic River Rail-road Company, for a road from the north line of the state, along the valleys of those streams, to the boundary of Massachusetts. Capital $2,000,000. To be completed in fifteen years, or forfeit charter.

The Rutland and Connecticut River Rail-road Company, for a road from Rutland, through Ludlow and Cavendish, to the Connecticut. Capital $500,000. To be completed in ten years, or forfeit charter.

The Brattleboro and Bennington Rail-road Company, for a rail or M'Adamized road between those towns. Capital $500,000. To be completed in ten years.

The Vermont Central Rail-road Company, for a rail-road through the Onion river valley to a point on the Connecticut to be determined hereafter. Capital $1,000,000. To be completed in twenty years, or forfeit charter.

Norwich and Hartford Forwarding Rail-road, incorporated in 1836, extends from the falls in the Connecticut near Hanover bridge, to Lyman's bridge. Capital $300,000.

MASSACHUSETTS.

RAIL-ROADS.

Eastern Rail-road, commenced July 22, 1836, and completed in 1839, extends from Boston in a north-eastern direction, to the division line between Massachusetts and New Hampshire, whence it is continued by another company to Portsmouth, in the last mentioned state. Length about 40 miles. The road passes through Lynn, Salem, Newburyport, &c., and is conducted under a part of the city of Salem, through a tunnel. It forms a part of the contemplated north-eastern line through New Hampshire and Maine, which sooner or later will reach the extreme eastern confines of Maine. On the fourth of July, 1839, there were upwards of 7000 passengers conveyed on this road, between Boston and Salem. An additional section extending from Newburyport to Portsmouth, will be opened for use in the course of the present year, (1840.) See New Hampshire.

Boston and Lowell Rail-road. The company under whose direction this road was constructed, was incorporated on the 5th June, 1830; its execution was commenced on the 28th November, 1831, and in June, 1835, it was opened for public use. Length 26.50 miles; 18 viaducts, one of them 1600 feet in length; 51 farm and road bridges; 12 street or road crossings; maximum grade, 1 in 528, or 10 feet per mile; least radius of curvature, 3000 feet. $\frac{24}{59}$ of the road consists of curved line, and $\frac{30}{59}$ of straight line; summit 125 feet above high tide, and the northern terminus of the road at Lowell is 94 feet above tide. The deep cut through which the road enters Lowell, deserves attention; for the distance of nearly a thousand feet, the solid rock has been excavated to the mean depth of forty feet, forming an immense chasm sixty feet wide at top, and about forty at bottom.

Plan of construction.—The first track leading from Boston to Lowell, 26 miles, is laid with the fish-bellied edge rail, of 35 lbs. per yard, and precisely of the pattern of the rail first employed on the Liverpool and Manchester Rail-road, and resting in cast iron chairs, supported on stone blocks and stone cross-sills alternately ; the bearings being three feet apart from centre to centre lengthwise of the road ; the blocks and sills being supported the whole length of the track, under each side, by a wall of dry rubble masonry, 3 feet in depth, 2½ feet wide at bottom, and 2 feet at top. Between these walls, and also outside of them, the clay or other material composing the road bed, and most convenient to be obtained, was used as a filling ; the contact of which, with the sides of the walls, and with so much of the under, and also the vertical, sides of the cross-sills, as did not rest upon the walls, caused both to be heaved by the frost during its action upon the clay and earthy matter. Great derangement of the track, with frequent fractures of the cross sills, was the consequence.

The trench walls, therefore, being found not to answer the preserving the first track from the effects of frost, as intended, were not resorted to in the laying of the part of the second track since put down. The fish bellied rail being also found to have a disadvantageous form, and being too weak for railway machinery of the most modern and economical weight, has not been used in the second track, about ten miles of which, from Boston to the water station, where the trains pass each other, have been laid.

In the second track the H rail, in 15 feet lengths, with square ends, weighing 55 lbs. per yard, being precisely the pattern of the Boston and Providence rail, is employed. It is laid upon stone blocks and stone cross sills, alternating, as above mentioned, for a part of the distance, whilst, in a considerable portion of the line, the stone sills extending across the track, are exclusively used. It is now preferred that every stone support to the rail should be a cross tie, and *blocks* will probably be rejected in the further construction of the second track. In the intermediate spaces, as well as at the joinings, the rails rest upon the stone dressed smooth, but without the intervention of a chair, or other material : at the joinings of the rail, however, they are let into the stone by an appropriate cut, so as to prevent a lateral movement of the rail, where it is the weakest.

The usual form of spike, with a head projecting on one side only, is used to hold down the rail by overlapping its base on each side, the spikes being driven into holes drilled into the stone, and filled with wooden plugs.

The sills and blocks rest upon a bed of gravel or sand, which fills a trench 7 feet wide and 3 feet deep, underlaying the entire track and well compacted by large rollers, (old mill stones were used as rollers.) The sills are 6 feet long, of a square section, 6 × 12, or 8 × 10, averaging about 10 × 10 inches. The cost of each sill delivered is stated to be $1 50. The cost per lineal rod, of filling the trench (exclusive of excavating it,) with the sand and gravel, compacting it with the rollers, and laying down the track upon the same, is said to be $5, the gravel or sand being supposed delivered in readiness alongside of the road.

The whole amount expended up to the 30th November, 1836, was for the road, - - - $1,323,522 00
For depots and aqueduct, - - - 79,895 67
For Engines and Cars, - - - 102,227 56

Amounting to $1,505,645 23

The repairs of rail-road in the year ending the 31st May 1838, are stated to have been, $15,340 69, which includes expenditures in improving the drainage of the road bed. The annual cost of a track alone of 26 miles in length as now laid upon this road with the H rail, would be $6,000, which is $230 77 per mile of single track.

At Lowell the road unites with rail-road to Nashua, and thus opens an uninterrupted rail-road communication with the populous manufacturing districts of Middlesex and Hillsboro counties. Cost about $1,650,000. A branch of this road leaves the main line in the town of Wilmington, and extends to Andover.

ANDOVER AND WILMINGTON RAIL-ROAD. This road branches off from the Boston and Lowell Rail-road, in Wilmington, and proceeds in a northern direction to Andover, distant 7.75 miles.

The company was incorporated in 1833, and in 1836 the road was opened for use.

ANDOVER AND HAVERHILL RAIL-ROAD, is an extension of the preceding road, which is now in active use. Length as extended 17.75 miles.

CHARLESTOWN BRANCH RAIL-ROAD, a branch of the Boston and Lowell Rail-road, in the town of Charlestown, extending to Gray's wharf, one mile.

BOSTON AND WORCESTER RAIL-ROAD, commences at the depot in Lincoln street, in the city of Boston; crosses the estuary of Charles River, and proceeds along the right bank of that stream to Newton; it thence continues, and at a distance of two miles from Newton, the line is conducted to the opposite side of the river into the town of Needham. Passing on through Natick and Westboro, it descends into the valley of Elizabeth river, which it ascends, mounts to the summit at Cutler's peak, and after traversing a broken and rough country for a considerable distance, descends the valley of Blackstone river, and enters the town of Worcester. Length 44 miles. Preparatory arrangements were made in 1831, and in the latter part of that year the work was commenced. Most of the work on this road is heavy, with much deep cutting and high embankments. At the crossing of Charles river is a costly construction of masonry and trestle work, and a short distance beyond is an embankment 680 feet in length, and 30 feet high, which is immediately succeeded by a cut 500 feet long and 31 feet deep, through granite rock.

Before the line reaches Natick, another high embankment occurs, and then a cut similar to the one just mentioned. About five miles eastward of Worcester, the road attains its greatest altitude, about 550 feet above high tide. The apex of the dividing ridge is here excavated to the depth of 37 feet for a distance of more than a quarter of a mile, through a hard slaty formation. About $\frac{16}{44}$ of the line are level; and the remainder has an average grade of 23 feet, and a maximum grade 30 feet per mile; least radius of curvature, 954 feet. Cost $1,700,000.

Plan of construction.—The rail employed is of the T pattern, weighing 38½ lbs. to the yard, in bars of fifteen feet in length; with scarfed, or oblique ends, at an angle of about 57 degrees with the line of the rail, and supported upon cast iron chairs, weighing each 15 lbs., the bottom of the vertical stem of the rail, resting on the bottom of the channel in the chair. The rails are tightened in the chair by two wrought iron keys, driven on the side of the rail outside of the track. The chair rests on cedar sleepers 7 feet long and 5 inches square, acting as cross ties, 3 feet apart from centre to centre. The sleepers cost 20 cents apiece delivered at Boston, and they are laid without any sill under them, but they rest upon a prism of rubble and broken stone 15 inches wide by about $2\frac{1}{2}$ feet in depth, laid in a trench dug for that purpose longitudinally of the road under each line of rails. This method of construction, however, having proved unsuccessful, the same company in their construction of the Millbury branch used an under-sill of chesnut, 8×3, and from 15 to 25 feet long; the joinings of these sills being supported by a piece of the same material from 3 to 4 feet long. The iron rails are stated to have cost $50 per ton, delivered in Boston. The spikes used in fastening the chair down upon the sleepers or cross ties of wood, above mentioned, are very heavy, being $7\frac{1}{2}$ inches long, 11-16 square, and weighing about one pound each.

The total outlay, as given in round numbers, was $1,700,000 of which $250,000 were for real estate, right of way, depot, buildings and machinery; leaving for the cost of the 44 miles of road construction, with a single track, (the graduation being mostly wide enough for two tracks) $1,450,000.

The Boston and Worcester Rail-road forms an important part of the Great Western rail-road communication, which will soon connect the eastern coast of Massachusetts with the valley of the Hudson, and thence, by the numerous rail-roads of New York, to the great lakes and the navigable waters of the Mississippi.

MILLBURY BRANCH of the preceding, diverges from the main line and extends to Millbury, 8 miles.

WESTERN RAIL-ROAD, extends from Worcester, through Springfield, to the western line of the state, at West Stockbridge, where it connects with the Hudson and Berkshire Rail-road, from Stockbridge to the city of Hudson. At Worcester it unites with the Boston and Worcester Rail-road, and thus forms a continuous line from the latter to the valley of the Hudson. In its length of 116.87 miles, the road surmounts four principal summits, whose respective elevations, exclusive of the necessary cuts, are 908, 1432, 1072 and 980 feet above tide.

The eastern division, from Worcester to Springfield, 54.27 miles in length, was commenced in 1837, and opened for use on the 1st October, 1839. The western, from Springfield to the N. York state line, 62.60 miles, was commenced in June 1838, and is now rapidly verging towards completion.

Plan of construction.—The road generally is graded for a single track, but all the deep cuttings and high embankments in its entire length, as well as the bridges on the eastern division, are prepared for a double track. The cutting for a single track 20, and for a double track 30 feet, except in rock, when the latter is 26 feet. Fillings for a single track 16, and for a double track, 26 feet.

The width of the track is 4 feet $8\frac{1}{2}$ inches. Edge rail, parallel form, of the depth of $3\frac{1}{2}$ inches, base 4 inches; top or tread, exclusive of swells, 2 inches; length of rail 5 yards; weight 55 lb. per yard. The rails are laid upon chesnut sleepers, 7 feet in length, 7 inches in depth, and from 7 to 12 inches wide. The sleepers rest upon hemlock sills, laid longitudinally of the road, 8 inches wide, and three inches thick, 4 feet 10 inches apart from centre to centre, with four additional bearing pieces, each 3 feet in length, laid at the joints of the iron rails; the sleepers are laid 3 feet from centre to centre. The rails, which are laid upon cast iron chairs, are secured in the usual manner, by spikes driven into the sleepers, notches being cut at the ends of the rails, about $\frac{1}{8}$ of an inch longer than the spike-holders in the plate, by which the effects of the changes of temperature are obviated.

On the eastern division, the maximum inclination is 60 feet per mile, and the minimum radius of curvature 1146 feet.

On the western division the former is 78.98 feet per mile, and the latter is 1042 feet. Locomotive power is used exclusively on the finished portions of the road.

Total cost of the eastern division, including depots, travelling apparatus, land damages, &c., $1,972,985 97, or about $36,135 per mile.

The estimated cost of the western division, with the buildings, &c. similar to the preceding, is $2,326,442 61.

WEST STOCKBRIDGE RAIL-ROAD, see New York.

ALBANY AND WEST STOCKBRIDGE RAIL-ROAD, see New York.

BERKSHIRE RAIL-ROAD, commences at Sheffield on the south boundary of the state, traverses the Housatonic valley and terminates at West Stockbridge, where it unites with the Great Western Rail-road from Worcester. Length about 25 miles.

BOSTON AND PROVIDENCE RAIL-ROAD, commences at Boston, passes through the towns of Roxbury, Dedham, Walpole, Foxboro and across Sekonk plains and Cove, to India bridge in Providence. This road, the work of a joint stock company incorporated in 1831, was opened for public use in June, 1835. Length 41 miles; single track, with side lines and turnouts. The line is arranged for two tracks, being 26 feet wide, and nearly straight. Cost $1,782,000.

The curvatures upon this road are very gentle, the least radius being 5730 feet. The highest grade ascending in a direction towards Boston is at the rate of 25 feet per mile, and in the opposite direction $37\frac{1}{2}$ feet per mile. The highest point upon the road is at Sharon, where it is 10 feet below the natural summit, and 256 feet above the ocean level. At Canton is the granite viaduct, of 700 feet in length, and upwards of 60 feet in height, across the valley of the Neponset river; and besides this noble structure, there are upon the road 1200 feet in length of wooden bridging, having spans of from 30 to 125 feet. There are likewise deep and long excavations and embankments, the former of which were very costly, on account of the presence of rock of the hardest description.

Plan of construction.—The rail-road is formed with a H rail in bars of 15 feet in length, with square joinings: 55 lbs. per yard. The chair is of cast iron, weighing about 10 lbs. The chairs are used only at the joinings of the rails, and the mode attempted to check an endwise movement of the rail consists in the narrowing of the sunk part of the top of the chair to a less width than that of the base, or lower web of the rail, and of cutting off a portion of that web on each side; thus narrowing the base of the rail correspondingly with the contracted opening in the upper part of the chair, so that the base of the rail so reduced in width will drop into the chair, and leave two small shoulders in the base of the rail that abut against the ends of the chair. The chair is let into the sleeper to prevent its moving laterally, whilst a longitudinal movement is resisted by the lateral strength of the 4 spikes, which are driven, two on each side, through holes in the chair, into the wood beneath. These spikes have brad heads, which overlap the lower web of the rails; and hence the plan is, that the adhesion of the spikes to the wood of the sleepers, or cross ties, in aid of the gravity of the materials, is relied upon to hold the rail down in the chair, and the latter firmly upon the wood. Upon the sleepers intermediate to those supporting the chairs, the rail is fastened with the brad spike, four in each sleeper, as is the case generally where the H rail is laid. The spikes used upon this road are ½ an inch square, 6 inches long, and weigh 9 oz. each. The rails rest on cross ties of white cedar, 3 feet apart from centre to centre, laid upon the bed of the road, except in a few instances on embankments, where under-sills were adopted to sustain the cross ties in proper line. The surface of the road bed generally presents a good material, being mostly composed of sand and gravel, yet these materials have not prevented injurious and partial settlings of the sleepers.

The cost of the repairs of road during the year 1836 upon the whole line of 41 miles, amounted to about $15,000, principally for keeping up the rail-way track.

This road forms a part of the route to New York by steamboats, which run between that city and Providence. It is also connected with another route from Boston to New York, via Providence and Stonington in Connecticut.

DEDHAM BRANCH RAIL-ROAD, an arm of the preceding, two miles in length, extends to Dedham.

TAUNTON BRANCH RAIL-ROAD, leaves the Boston and Providence Rail-road in Mansfield, and proceeds to Taunton, 11 miles. An extension of this line to New Bedford, is now in progress, 24 miles in length.

NEW BEDFORD AND FALL RIVER RAIL-ROAD, extends between those places, Massachusetts section 8 miles—Rhode Island section, 5 miles; total length, 13 miles.

NORWICH AND WORCESTER RAIL-ROAD, see Connecticut.

PITTSFIELD AND WEST STOCKBRIDGE RAIL-ROAD, about 15 miles.

SEKONK RAIL-ROAD, from Providence to Sekonk, Bristol county, Mass. Length, about 5 miles.

QUINCY RAIL-ROAD, in Norfolk county, is used for transporting granite from the quarries in the town of Quincy to the landing on Neponset river. Length 3 miles; branches 1 mile; one inclined plane, 275 feet in length, single track.

CANALS.

MIDDLESEX CANAL. The company was incorporated in 1789, and their canal was not completed till the year 1808; nearly nineteen years having elapsed from its commencement to its completion. It extends from Chelmsford, on the Merimac, two miles above Lowell, to one of the inlets of Charles river, in Charlestown. This canal, like the Boston and Lowell Rail-road, is designed to facilitate the intercourse between the Merimac valley, in New Hampshire, and Boston. It is about 30 miles in length, and in a N. W. direction from Boston; 30 feet wide at top, 20 at bottom, and 3 feet deep. Rise from Boston 104; fall towards Chelmsford 32 feet; total lockage, 136 feet; 20 locks; cost $528.000.

PAWTUCKET CANAL, in the town of Lowell, is used both for navigation and manufacturing purposes. Length 1.50 miles; 90 feet wide at top, and 4 feet deep; lockage 32 feet.

BLACKSTONE CANAL commences at Providence, R. I., and pursues nearly a due north course, and enters the valley of the Pawtucket river, which it follows to the town of Worcester. Length 45 miles; depth 4 feet; 34 feet wide at top; 18 at bottom; 48 locks, 80 by 10 feet; cost $600,000; completed in 1828.

HAMPSHIRE AND HAMPDEN CANAL, is a prolongation of the Farmington Canal of Connecticut. It commences at the point of termination of that work in Southwick, Hampden county, and proceeding through Westfield and Easthampton, and encircling the base of Mount Tom, terminates at Northampton, in Hampshire county. Length 22 miles; 36 feet wide at top, 20 at bottom, and 4 feet deep; locks 80 by 12 feet; commenced in 1825; finished in 1831.

MONTAGUE CANAL, near the Montague falls, in Connecticut river. 3 miles long; 25 feet wide; 3 feet deep; 8 locks; lockage 75 feet.

South Hadley Canal, along the falls of the Connecticut, in the town of South Hadley. Length 2 miles; 5 locks.

Aggregate length of rail-roads in Massachusetts, 407.31 miles.
" " canals, " 79.50 "

RHODE ISLAND.

RAIL-ROADS.

Providence and Stonington Rail-road. This rail-road was intended to open a direct communication between the cities of New York and Boston. Long Island Sound, which is sufficiently free from ice during the winter, to admit of its uninterrupted navigation at all seasons of the year, forms a part of this route, as well as that by the way of Narraganset Bay; though Narraganset Bay is avoided by the Stonington route, yet most travellers prefer the former.

The latter rail-road commences at Providence, where it connects with the Boston and Providence rail-road, by a steam ferry boat on the Providence river, proceeds in nearly a straight line through the towns of Cranston, Warwick, and East Greenwich, to Hurst's Run, where it deflects towards the southwest, and continues that course to Sherman's Pond, in South Kingston; here the road enters the valley of Charles river, which is frequently intersected by it, passing alternately along its right and left banks, until it reaches Paquent Run, the outlet of Watchaug Pond, and thence over the high ground of Westerly, and across the Pawcatuck river, and terminates at the cove of Stonington, on Long Island Sound, in the state of Connecticut.

RHODE ISLAND.

Plan of Construction—The rail is of the H pattern, of 58 lbs. per yard, in bars of 15 feet in length, with square ends; that is, the top surfaces of the bars join in a direction at right angles with the lengthwise line of the rail. It is very similar in form to the rail of 55 lbs. per yard upon the Boston and Providence road. The cast iron chair supporting the ends of the rails, is nearly similar in form and weight to that on the Boston and Providence road, (about 10 lbs.) as are also the spikes of 6 inches long, $\frac{1}{2}$ an inch square in the shank, and weighing 9 ounces each. The sleepers are laid 3 feet apart from centre to centre, are of white cedar from Maine, hewn top and bottom to a uniform thickness of 6 inches, 7 feet long, and were delivered at various landings on or near the line of the road, at 26 cents a piece, dressed. These sleepers rest on under sills of hemlock, 3×8 inches in the cuts, and 3×10 upon the embankments, and of a length from 20 to 30 feet, and costing upon the landing at the rate of $13 75 per M. Under every joining of two adjacent under sill pieces, is placed a short sill of the same scantling, 5 feet long, to support the joint; and under every sleeper on which two rails join, there are placed on each side of the under sill, and close to it, the half of a sleeper, each consequently $3\frac{1}{2}$ feet in length. This superstructure reposes on a bed of sand or gravel, with which the entire surface of the road-bed is covered, to a depth of 18 inches. The gravelly nature of the soil throughout a great part of the route, made this *ballasting* not very expensive.

The cost of transporting all the materials of iron and wood composing the track, from the landings to the line, an average distance of 5 miles, was $1 75 per rod, in length of the road; the cost of laying the track $2 25 per rod; and that of filling it with gravel between the sleepers, after it was laid, 25 cents per rod; making the whole cost of laying the track $4 25 per rod.

Length 47 miles; company chartered in 1832; work commenced in 1835; completed in 1837; summit 302 feet above tide; graduation 14 miles level or nearly so; maximum grade 33 feet per mile; average of the remainder, about 13 feet per mile; minimum radius of curvature, 1,637 feet, except a short distance near Providence, where it is 480 feet; single track, but is graded for a double track; cost about $2,000,000.

New Bedford and Fall River Rail-road. See Massachusetts.

CANALS.

Blackstone Canal. See Massachusetts.

Aggregate length of rail-roads in Rhode Island, 47 miles.
" " canals " 38 "

CONNECTICUT.

RAIL-ROADS.

Norwich and Worcester Rail-road. On leaving the steam-boat wharf at Norwich, the road pursues a general north-east course, along the right bank of Quinebaug river, to Danielsonville, where it crosses the main stream, and enters the valley if Five Mile Run. From its crossing in Killingly, the road proceeds nearly due north, leaves Five Mile Run, and regains the bank of the Quinebaug, which is followed to the mouth of French Creek, thence up that creek, it crosses the state line, and enters Massachusetts. Continuing along the left bank of French River, and passing through the towns of Webster, Milbury and Auburn, it enters that of Worcester, where it meets the rail-road from Boston, on the east, and that to Springfield, &c., on the west. Length 58.50 miles; maximum grade 20 feet per mile; average inclination 11 feet per mile; cost about $1,000,000. This road, in connection with the steam-boats on Long Island Sound, and the Boston and Worcester rail-road, will furnish a desirable route between New York and Boston, and afford important accommodations to the densely populated and manufacturing section of country through which it passes.

New Haven and Hartford Rail-road, commencing near Mill Creek, in New Haven, and crossing that creek, the road is conducted to the bank of the Quinnipiack, in North Haven. Here the road crosses the stream by an aqueduct, and gains its left bank, which is ascended to Meriden, and thence through the towns of Berlin and Weathersfield, to Hartford, its present point of termination. Length 40 miles. It is proposed to extend this road to Springfield, in Massachusetts, a further distance of 28 miles.

HOUSATONIC RAIL-ROAD. The charter, under the authority of which this company is organized, was granted by the legislature of Connecticut, in May, 1836, with the powers usually conferred upon such bodies, and giving them authority to construct a rail-road from the north line of the state, adjoining the town of Sheffield, in Massachusetts, down the valley of the Housatonic, by New Milford, to the town of Brookfield, and thence to the city of Bridgeport, in Fairfield county.

Thus it will be perceived that the Housatonic Rail-road opens a communication, not only with the western towns of Connecticut, but also with those of Berkshire, in Massachusetts, and will, ere long, be intersected by the Berkshire Rail-road, uniting the Housatonic works with the Hudson and West Stockbridge Rail-road, now in successful operation; and also with the Albany and West Stockbridge Rail-road. Length of the Housatonic Rail-road, 73 miles; commenced in July, 1837; it has one tunnel, and an embankment 70 feet high; maximum grade 40 feet per mile; minimum curviture 1,000 feet radius; cost about $1,000,000, or $13,700 per mile.

BRIDGEPORT AND SAWPITTS RAIL-ROAD, in Fairfield county. This road commences in Bridgeport, pursues nearly a west course, and passes into the village of Fairfield; thence in the same direction to Southport and Saugatuck river, which the line crosses about one and a half miles south of Westport; thence in a direct line to Norwalk river at Old Well; here the line deflects towards the southwest, which course is maintained in ascending a ridge between the latter river and Five Mile Run; at this ridge an excavation of 42 feet occurs; from this point the line continues to the Short Rocks, and over Good Wife's and Noroton rivers, to Stamford; thence by Miamus river, Put's ridge, and Bynum's valley, it crosses into the village of Sawpitts, in Westchester county, New York. Length 30.46 miles; maximum grade 40 feet per mile; cost $506,457 96, or $16,628 per mile.

Rail-roads from Hartford to Springfield—from Worcester to Hartford—in Fairfield county, &c., are in progress.

CANALS.

FARMINGTON CANAL, commences at New Haven, passes

along the valleys of Mill creek and Farmington river, and intersects the north line of the state in the town of Suffield, where it unites with the Hampshire and Hampden Canal of Massachusetts. Length 56 miles; general course nearly north; 36 feet wide at top; 20 at bottom; and 4 feet deep; locks, 80 by 12 feet in the clear; cost $600,000.

ENFIELD CANAL is designed to overcome the Enfield falls, in Connecticut river. Length 5.50 miles; 3 locks each of 10 feet lift, 90 by 20 feet.

Aggregate length of rail-roads in Connecticut, 188.46 miles.
      "   "   canals   61.50   "

NEW YORK.

CANALS.

The project of uniting the Hudson with the western lakes by an "artificial river," appears to have engaged the attention of the citizens of New York from the earliest periods of its settlement. No one, indeed, who had studied the physical aspect of the region through which the Erie Canal now passes, could fail to observe its peculiar fitness for such a work; and as the population increased in the western division of the state, its importance became more obvious, and its practicability demonstrated by a more thorough acquaintance with the physical structure of that part of the country.

The provincial legislature so early as 1768, had its attention drawn to the subject by the then governor of the province, and from this period it formed a prominent theme for its deliberation. The political state of the times which preceded the revolutionary struggle, and the ensuing conflict, arrested for a time, all further proceeding in relation to this great enterprise, until after the close of the war; when efforts for its accomplishment, were renewed by its advocates.

Although the project had been frequently brought before the state legislature, no legal steps had been taken to secure its execution, until 1808, when the assembly appointed a committee to investigate the subject, and if found practicable, to solicit the co-operation of the general government in " the accomplishment of the great work."

The result of this investigation was highly favourable, and established, beyond doubt, the entire feasibility of the proposed measure, though nothing further was done at that time. Notwithstanding the failure on the part of the legislature to act promptly upon the report of its committee, it had the effect of arousing the citizens generally to a just appreciation of its importance, and the ranks of its advocates rapidly augmented in numbers and influence. In 1810 the legislature appointed commissioners to examine and survey the entire route from the Hudson to Lake Erie; this examination resulted in a recommendation, on the part of the commissioners, that immediate efforts be made to induce the general government to adopt and execute the proposed connection. This proposition, as well as subsequent applications to the various state governments for aid, proving entirely fruitless, the state was, fortunately, thrown upon her own resources, and thus escaped the trammels of those entangling alliances that must have proved a source of endless and vexatious embarrassment to the conductors of the public works.

Undismayed by these untoward events, the friends of the system advanced with a steady pace until checked by the war with England, which led to the dissolution of the board of canal commissioners, and the consequent suspension of all proceedings with regard to the object of its appointment.

Soon after the cessation of hostilities between the United States and Great Britain, in 1815, the subject of internal improvement was revived and zealously pressed upon public attention; meetings were held in the city of New York and elsewhere, which resulted in memorializing the legislature in favour of the proposed improvements: and in 1816, a board of commissioners was again appointed with powers similar to those exercised by the board of 1812.

The new board displayed such energy and promptitude in the performance of its duties, that the legislature was enabled to act definitively; and on the 15th of April, 1817, was passed that memorable law upon which the system of internal improvement in New York, is based. Contracts were immediately entered into, and on the 4th of July, 1817, the execution of the Erie Canal was commenced in the neighbourhood of Rome; and in rather more than two years, that portion of the work extending from Utica to Montezuma, was opened for navigation. The other sections continued to advance; portions of each were opened for use from time to time until October, 1825, when the whole line was ready for use. The completion of this magnificent work was celebrated with great pomp and parade; an event so well calculated to inspire feelings of exultation on the part of its friends, fully justified the measures then adopted to perpetuate its remembrance; which, though regarded by some as extravagant, were viewed by all as an excusable ebulition of feeling, on the consummation of their favourite object.

The moral effects of this herculean achievement, are now visible in every direction. Stimulated by the complete success of the New York system, other states have essayed to imitate an example fraught with such incalculable benefits. Canals and rail-roads now abound every where; and every where may be seen preparations for augmenting the number.

ERIE CANAL. Among the former, the Erie Canal still maintains its supremacy: it is unquestionably the first in point of length, and by far the most important canal in the United States. Its general course from Albany is a little north of west. Leaving Albany it passes along the right banks of the Hudson and Mohawk, crossing the latter at Middletown; following the left or north bank of the Mohawk about 12 miles, it re-crosses that river over what is termed the "upper aqueduct;" pursuing the south bank of the Mohawk through Schenectady, Schoharie, Canajoharie, and Little Falls village, it enters the flourishing town of Utica, 108 miles by the canal from Albany. Continuing its course along the southern bank of the Mohawk by Whitesboro, Rome, Lenox, Syracuse, Jordan, Montezuma, Lyons, Palmyra, Pittsford, to Rochester (distant 160 miles from Utica), where it crosses the Genesee by a splendid aqueduct, 804 feet long, built of hewn stone, and supported by eleven arches. From Rochester the canal assumes a more western direction until it reaches Lockport, after passing the towns of Brockport, Albion, Middleport, and some other "ports" of lesser note; distance from Rochester to Lockport 63 miles. At Lockport the canal ascends the mountain ridge, by five double combined locks, each 12.4 feet rise. Nine miles from Lockport, the canal enters Tonnewanda creek, at the little village of Pendleton, from which to Tonnewanda village, situated at the mouth of the creek and distant from the former about 10 miles, the canal is identified with the Tonnewanda. At a further distance of 12 miles, this magnificent work terminates, at the city of Buffalo. Entire length of the Erie Canal 363 miles;

NEW YORK.

40 feet wide at top, 28 at bottom, and 4 feet deep; 84 locks on the main line, each 90 by 15 feet; total lockage 688 feet; 8 large feeders; 18 aqueducts; declivity from Buffalo to Rochester 4 feet; rise 630 feet; fall 62 feet; total rise and fall 692 feet.

Among the aqueducts which cross the Mohawk, is one 1188 feet in length.

What is called the "long level," is a portion of the canal $69\frac{1}{2}$ miles long, without any intervening lock. It begins at Frankfort, 8 miles east of Utica, and terminates at Syracuse.

The great embankment, 72 feet high, is situated between Palmyra and Pittsford, about 255 miles west of Albany. Cost $10,731,595. An extension of the Erie Canal from Buffalo, in the direction of Warren, Pa., is proposed.

In 1835, a project for enlarging the Erie Canal was adopted. It arose from the necessity of repairing some parts of the canal, especially the locks and aqueducts. The want of additional facilities for conducting the increased trade had become apparent, when it was determined to augment the capacity of the existing canal rather than construct another line, which had been long contemplated. Considerable progress has been made in the work. The expense of the enlargement is now estimated at $23,402,863; of which $4,182,565 have already been paid, and contracts made on the work amounting in all to $10,683,565. It is proposed to widen the canal to 60 or 70 feet, and to deepen it 2 feet. If completed throughout on the scale adopted by the canal commissioners, the canal will surpass in magnitude every other work of internal improvement.

The aqueduct now being constructed across the Genesee river at Rochester, which is to be a splendid structure, will, according to the estimate, cost $422,245 28. Including the wings, it is 858 feet long and 28 feet in height from the base of the piers to the top of the parapet. It has 7 arches of 52 feet span, 6 piers and 2 abutments, each 10 feet thick. The aqueduct is 75 feet 6 inches in width at the base of the piers, and 67 feet 8 inches at the top of coping. The width of the trunk, in the clear, will be 45 feet, affording space for a double boatway.

The works at Lockport consist of five double combined locks, of the enlarged size, having a total lift of 55 feet and 9 inches. The total cost of these locks, $558.433 04. The old locks, displaced by the new set, were built in 1823-'24-'25, and cost, exclusive of excavation, $123,309 55.

The length of the rock-cutting through the ridge at Lockport, is 2½ miles, commencing at the head of the locks, 62 feet wide, with vertical sides; and it is deepened 3½ feet; the cost of enlarging this section of the canal will be $762,635 07.

With the exception of the Hudson and Delaware Canal, all the minor canals in New York, may be regarded as branches of the Erie Canal or main trunk. They are as follows:— Champlain Canal; Chenango Canal; Black River Canal; Oswego Canal; Cayuga and Seneca Canal; Crooked Lake Canal; Chemung Canal, and Genesee Valley Canal. The entire cost of the finished canals belonging to the state, is $12,072,032 25.

CHAMPLAIN CANAL. From its junction with the Erie Canal, nine miles from Albany, the Champlain Canal pursues its course through Waterford, Stillwater, and Bemus' Heights, along the right bank of the Hudson, to within three miles of Fort Miller where it takes the river for three miles. At Fort Miller Falls is a canal of half a mile in length, and then again it takes the river 8 miles to Fort Edward, from which place the canal pursues a north-east course to Whitehall, after passing the villages of Sandy Hill and Fort Ann. Length of the Champlain Canal from its junction with the Erie Canal to Whitehall, including the Glenns Falls branch, 11 miles in length, and river navigation, 76 miles; 40 feet wide at top, 28 feet at bottom, 4 feet deep; 21 locks, each 90 by 14 feet; rise 134, fall 54 feet; total lockage 188 feet: elevation of Lake Champlain above the Erie Canal, at the junction, 38½ feet; commenced in 1816; completed in 1819; cost $1,179,872.

It has a lateral cut, at Waterford, connecting it with the Hudson. At the upper part of the city of Troy, 3 miles below Waterford, is constructed a *state dam* and *lock*, which forms a capacious basin.

The annual net receipts on the Champlain and Erie Canal are about $1,000,000, and the average cost for repair is about $700 per mile.

NEW YORK. 55

Statement of the Revenue from Tolls of the Erie and Champlain Canals, the Expenditure for maintaining them, with the Surplus of each year, from 1826 *to Sept.* 30th, 1839, *according to the Comptroller's Report.*

Year.	Revenue.	Expenditure.	Surplus.
1826,	$715,245.89	$579,667.57	$135,578.32
1827,	846,651.73	446,293.76	400,357.97
1828,	794,054.25	312,377.27	481,676.98
1829,	771,012.85	292,318.71	478,694.14
1830,	1,005,392.32	236,972.97	768,419.35
1831, 9mo.	708,426.42	172,408.80	536,017.62
1832,	1,055,027.88	374,231.10	680,796.78
1833,	1,317,464.33	375,147.52	942,316.81
1834,	1,314,799.69	448,775.82	866,023.87
1835,	1,433,456.38	463,420.18	970,036.20
1836,	1,551,057.18	425,539.39	1,125,517.79
1837,	1,274,403.94	477,182.88	797,221.06
1838,	1,415,279.79	513,279.99	901,999.70
1839,	1,617,246.00	505,729.65	1,111,517.00

It was in the year 1826, that the tolls began to exceed the expenditure; and the whole debt for these canals at the close of the year 1826, exclusive of the interest of that year, amounted to $10,272,316 75. According to the computation of the Comptroller, the surplus revenue had, in 1838, paid the interest on the debt, and reduced the amount of the debt to the sum of $8,459,069 17.

CHENANGO CANAL leaves the Erie Canal at Utica, and proceeds in a south-west direction, over Paris Hill, and enters the valley of Chenango river, which is followed to Binghamton, at its junction with the Susquehanna. In its course the canal passes through Oneida, Madison, Chenango and Broome counties, and the towns of New Hartford, Clinton, Hamilton, Sherburn, Norwich and Chenango Forks. Length 97 miles; rise from Utica, 706 feet; fall from summit to Binghamton, 303 feet;

total lockage, 1,009 feet; 116 lift and 1 guard locks, 5 of stone, and the remainder of "composite," (stone faced with timber;) 19 aqueducts; 52 culverts; 21 waste weirs; 56 road, and 106 farm bridges; 53 feeder bridges; 12 dams; 17 miles of feeders; cost, according to the canal commissioners' report, $1,737,703, 22; commenced in 1833, and completed in 1838. The summit level of Chenango Canal is supplied with water from 7 reservoirs constructed for that purpose.

BLACK RIVER CANAL consists of a succession of canals and slack water pools. It commences on the Erie Canal, at Rome, in Oneida, county, by a canal of 36 miles in length, which terminates at the High Falls of Black River, Lewis county; thence to Carthage, in Jefferson county; the river, the navigation of which has been improved, completes the line. Length of navigation, natural and artificial, including a navigable feeder of 9 miles, extending to Boonville, 85 miles; ascent and descent from Rome to Carthage, 1,078 feet; cost of the whole, $2,141,601 63.

The various surveys directed by the legislature, with a view to the improvement of the northern tributaries of the Hudson, have been successfully prosecuted, and will, no doubt, result in the adoption of measures to that effect.

OSWEGO CANAL. This, like the Black River navigation, consists of pools and canals. It commences on the Erie Canal, near Syracuse, in Onondaga county, passes along the valley of, and nearly parallel to, the Osweago river into Lake Ontario, through the villages of Liverpool, Three River Point, Oswego Falls, and Oswego, on the right bank of the river; general course from Syracuse, northwest; length 38 miles; 14 locks of stone, and 6 guard locks, each 17 by 90 feet; descent 123 feet; cost $525,115; commenced in 1826, and completed in 1828.

CAYUGA AND SENECA CANAL (pools and canal,) connects the Seneca and Cayuga lakes with the Erie Canal, which it leaves at Montezuma, passing through Waterloo, the seat of justice of Seneca county, and along Seneca outlet, to Geneva; a branch, 2 miles in length, leaves the main line, proceeds to East Cayuga; course from Montezuma, southwest; length, including branch, 23 miles; 11 locks; descent 73 feet; commenced in 1827, and completed in 1829; cost $214,000. By means of

NEW YORK. 57

this canal a communication is afforded between Cayuga and Seneca lakes and the Erie Canal.

CROOKED LAKE CANAL, from Pennyan to the outlet of Seneca lake. Length 7.75 miles; lockage 269 feet; 27 lift and one guard locks, built of wood; 3 culverts; 12 bridges; cost $137,000.

CHEMUNG CANAL, extends from the head waters of Seneca lake to Elmira, on the Tioga branch of the Susquehanna. Length 23 miles; with a navigable feeder of 16 miles, from Knoxville. It has one guard, and 52 lift locks, of wood, overcoming an ascent and descent of 516 feet; 3 aqueducts; 5 culverts; 76 bridges; cost $344,000. Measures have been adopted by the state authorities, to extend this canal to the Pennsylvania line, near Tioga Point. The rail-road from the Blossburg coal mines, Pennsylvania, will join the western termination of the Chemung Canal, at Knoxville. Commenced in 1830, and completed in 1833.

The revenue derived from the minor canals is inadequate to the payment of interest on the loans contracted for their construction. The current expenses of, together with the interest on the loans contracted for those canals, have hitherto exceeded the revenue derived from their use. The deficiency, which in 1839 amounted to $182,688 10, is a charge upon the treasury.

Whether the state canals will ever yield the current interest on the cost of their construction, is a question, however, of secondary consideration, compared with the more important commercial, agricultural and political improvement in the social condition of the people, which must follow their introduction.

CONEWANGO CANAL, from Buffalo to the Pennsylvania line, in the direction of Warren, Pa. Surveys have been made for this work under the authority of the state government.

UTICA AND OSWEGO SHIP CANAL. This work is to extend from Utica to Oswego, on Lake Ontario, about 90 miles in length, including a large portion of natural navigation, by the Oswego and Oneida rivers, and Oneida lake, which is adapted to vessels of a large class, and which will require but little improvement. The entire line, consisting of 35 miles of canal and 57 of river or lake navigation, will probably cost about

$1,200,000. The legislature, during the session of 1839, appropriated $75,000 to be expended in the improvement of the Oneida river.

DELAWARE AND HUDSON CANAL, unites the Hudson river with the Carbondale coal mines, in Pa. It commences on the left bank of the Walkill, at Eddyville, about 2 miles south of Kingston, and proceeds in a general southwest direction, along the valleys of the Walkill, Rondout, Butterkill and Nevisink rivers, and through Kingston, Marbletown, Mombacus, and Warwasing, in Ulster county, to Port Jervis, at Carpenter's Point, on the Delaware. At this point the canal deflects towards the northwest and pursues that course, along the left bank of the Delaware, to a dam near the mouth of the Lackawaxen creek; here the canal crosses the Delaware, and enters the valley of the Lackawaxen, which is thence followed along its north declivity, to Honesdale, where it terminates, and where the rail-road to Carbondale commences.

The rail-road section of this improvement, 16.50 miles in length, has a rise towards the summit of 912 feet, and thence to the coal mines it descends 850 feet, overcome by 7 inclined planes, with an inclination of one in twelve. Length of the New York section of the canal, 83 miles; Pennsylvania section, 25 miles; and rail-road, 16.50 miles; total length from the Hudson to Carbondale, 124.50 miles; rise from the Hudson to the summit, in Sullivan county, 535 feet; thence to Carpenter's Point is a descent of 80; thence to the crossing of the Delaware, a rise of 148; and thence to Honesdale, a rise of 187 feet; total lockage 950; Honesdale 870 feet above tide water. The canal is 4 feet deep, and varies in width from 32 to 36 feet; 107 locks, each 76 by 9 feet. These works were originally executed by two distinct companies; the Hudson and Delaware Canal Company, of New York, and the Lackawaxen Canal Company of Pennsylvania. These corporations having been united, now form one interest. By acts of the legislature of New York, the credit of that state was loaned to the company, for $800,000, for the redemption of which, the canal and its appendages are pledged. The former company was incorporated in 1823, with banking privileges. The work was commenced in 1825, and completed in 1829, at a cost of $1,875,000.

GENESEE VALLEY CANAL. From Rochester to Olean on the Allegany river; 119$\frac{50}{80}$ miles in length.

This canal pursues the valley of the Genesee, the head waters of which are elevated 1000 or 1500 feet above tide-water. After flowing with a gradual descent nearly through the county of Allegany, they are suddenly precipitated upon the alluvial bottom, through a succession of cataracts, and rapids, which extend northwardly for 17 or 18 miles from Portageville to, and terminate at, Mount Morris.

The Genesee Valley canal partakes of the irregularities which are occasioned by these abrupt transitions in the character of the river. For the first 36 miles, after leaving Rochester, it passes through the rich low land district of the Genesee valley, between the Erie canal and Mount Morris, and it attains the latter point by a lockage of only 95 feet divided into ten locks. But immediately on leaving Mount Morris, the character of the canal undergoes a sudden alteration. For nearly 17 miles south of that place, the bed of the river is confined within a precipitous and rocky defile, varying from one hundred to four hundred feet in depth, which has been abraided by the stream in its rapid descent from the elevated plateau at its source. This sudden ascent is overcome by very numerous locks, a tunnel and other expensive works. After surmounting this gorge, the canal finally attains its summit level. The summit is 11.50 miles long, and from its southern extremity, the canal descends for 10 miles down a gentle declivity to Olean. The summit level is 978 feet above the level of Lake Erie, and 1546 above tide-water. The lockage on the main canal is 1063 feet, exclusive of the Danville branch. Cost $4,900,000; 114 locks; 81 of stone, and 33 composite.

DANSVILLE BRANCH of the Genesee Valley Canal, extends from near Mount Morris to Dansville, 11 miles long. Cost $314,520 43; lockage 83 feet.

HARLEM CANAL, extends from the Hudson to the East River, across Manhatten or New York Island; length 3 miles; 60 feet wide, and from 6 to 7 deep; 2 tide locks; cost $550,000. This work remains in an unfinished condition.

CROTON AQUEDUCT, is designed for the supply of the city of New York with pure and wholesome water. Of the true character and magnitude of this important work, but few, even of the citizens of New York have an adequate conception. The great amount of mere manual labor in excavating and tunnelling through solid rock, and the mechanical skill required in the erection of the bridges, culverts, walls and other erections which go to complete the great work, can scarcely be understood by those who are not practically acquainted with such things.

The Croton Aqueduct is 40.56 miles in length. Its dimensions are about 6 feet at bottom, 7 feet at top, and from 8 to 10 feet in height. It is higher for the first 5 miles after leaving Croton than it is on the remainder of its route. The foundation of the aqueduct is stone, well laid, and the interstices filled up with rubble, and over this a bed of concrete, composed of cement, broken stone and gravel, in due proportions, well mixed and combined together, except where the earth is of a compact and dry consistence, when the stone foundation is omitted, and the bed of concrete laid on the earth foundation. The side walls are of good building stone, 39 inches thick at bottom, and 27 at top, having a batter of 3 inches by 12.

These walls are laid in regular courses, and built with great care, under strict inspection, in order that the water may be prevented from escaping or entering the aqueduct. The bottom or flooring of the aqueduct is an inverted arch, and the top or roof is a semi-circle. Both arches are turned with brick of the most durable quality, and the interior surface of the side walls, has a coat of hydraulic mortar, and is also lined with the same material of brick. The whole of the mason work is constructed with mortar composed of the best hydraulic lime and sand; and there can scarcely be a doubt, that the work will stand the test of time, and answer all the purposes for which it is designed.

The materials used are good building stone, of the proper degree of hardness and durability, free from all metals, particularly iron: gneis is preferred to any other, both because it is more plentiful, and more easily worked. Some limestone is also used, but not until it has the express permit of the Resident Engineer. Brick is the next material; it is required to be from the centre of the kiln, such as is thoroughly burnt, free from lime or any other impurity, and to possess a clear ringing sound when struck. The worst accepted are such as cost from $5 to $7 a thousand. Next is the cement, from which the concrete and masonry generally are formed. The commissioners' specifications are very explicit relative to the manufacture of this article, requiring that the name of the manufacturer should be known; that the cement shall not have been made more than six months before being used; that it shall be transported from the factory in water-tight casks; and in addition to all this, that each parcel or cargo received shall be thoroughly tested, either by officers appointed for the purpose, or by the Resident Engineer himself. These are the principal materials, stone, brick, and cement. The stone is required to be always clean, and in hot weather, kept wet, and when laid in the wall requiring mortar, it must " swim" in the cement—that is, when the stone is lifted up from its bed, no point or surface of the stone must touch the one below it, each stone must be *surrounded* by cement. When the weather is hot, the top of the wall must be kept moist, and in cold weather all the masonry must be covered so effectually, as to protect it perfectly. The brick must be laid true and even, allowing $\frac{3}{8}$ of an inch joint, or thereabouts. In hot weather, they are to be soaked in water, and to be kept wet while being laid. The cement is mixed in different proportions, according to the work required. For stone work, the proportions are one part of cement to three of sand, (the sand to be of medium size, sharp grained and clean—river sand is accepted.) For brickwork, the proportions are one of cement to two of sand; for concrete, one part of cement, three of sand, and three of clean building stone, broken about as fine as that used for Macadamizing. Concrete is used for forming artificial foundations, is mixed with as little water as possible, and when laid in any part of the work, is left undisturbed forty-eight hours; at the expiration of this time it has become so hard, that a blow with a pickaxe will not break it—it becomes quite a rock.

The aqueduct, maintaining a uniform descent, requires that in places the earth should be cut away, and in crossing vallies that they should be filled up. In the former case, the sides of the cut are left standing at a slope of one-half to one—that is, if the perpendicular height of the side of the cut be 6 feet, it will fall off from directly above its base 3 feet. It is one-half horizontal to one vertical. The base of the cut is always 13 feet wide. Pegs, showing the bottom of the side walls, and of the reversed arch in brick are given by the engineers, who, at the same time, determine the centres, if necessary, from these data. The builder lays a small layer of concrete, *at least* three inches, whose top shall be as high as the top of the peg just set—on this concrete he proceeds to build the side walls of the aqueduct. The side walls being done, they are filled in behind them, up to the top, with earth, to prevent strain or damage, also to act as a support, and cover up the work as fast as possible. Then the concrete is laid for the bottom of the reversed arch in brick, by means of moulds placed every ten feet apart. When thoroughly set, the brick work is commenced. Selecting the best brick (and it has all been most thoroughly inspected) the reversed arch is laid, and then the "brick-facing"—that is, facing the inside of the wall with brick, when carried up to the top of the wall. The upper arch, consisting of two ring courses (with occasional headers) is thrown; the arch is covered with a thick coating of plaster, and the angle made by the top of the wall and arch, filled with the same kind of masonry as the side walls.

You will perceive it to be a long brick vault stretching from New York to Croton—ascending at the rate of nearly 14 inches in a mile. The earth removed in the excavation is then "back-filled" over the aqueduct until it is 4 feet deep over the crown of the arch, level on top, and 10 or 8 feet wide, and the sides slope $1\frac{1}{2}$ to 1. When the ground is too steep, a "protection wall" is introduced, this is laid dry, *i. e.*, without mortar, and made to slope one half to one, or one to one, at an angle of 45°. So much for the aqueduct in "open cutting in earth." When a valley is crossed, a heavy wall fifteen feet wide on top, with sides sloping one-twelfth to one, is built with large stones firmly embedded in small broken ones. On the top of this wall, a foot of concrete is placed, the aqueduct, as usual, is built *on that.* As water passes through vallies, a stone passage way, called "culvert," is made of suitable dimensions.

The dam at Croton, about 5 miles above its mouth, will back the river several miles, and cover with water, exclusive of its present bed, between five and six hundred acres, and thus form the great reservoir, which will contain 100,000,000 of gallons for each foot in depth from the surface. It is a submarine mound, 100 feet in length; 70 feet wide at bottom, and 7 feet at top; with an average height of 40 feet; built of stone and hydraulic cement. Immediately after the aqueduct leaves the dam, it passes through the "Corporation Tunnel," 180 feet in length, and proceeding one mile further, it crosses Lounsberry brook, where there is a culvert of 6 feet in diameter, and 66 in length. In crossing this valley, the grade line is 40 feet above the brook, and 55 to the top of the aqueduct. Five miles from Lounsberry's, Indian Brook is crossed by a culvert, 8 feet in diameter, and 142 feet long; and the aqueduct is conducted through Benveneu tunnel, 720 feet long, and the Acker's Brook Tunnel, 166 feet long. Half a mile from Indian Brook occurs another, Hoagshill tunnel, 276 feet in length. From this to Sing Sing there are several small valleys crossed by the aqueduct, varying from 5 to 18 feet, and averaging about 12 feet each, below the grade line, and 25 feet below the top covering of the aqueduct. At Sing Sing it passes the Sing Sing Kill tunnel, 336 feet in length, cut through solid rock, and arrives at the crossing of Sing Sing Kill, 2 miles from Indian Brook. The chasm, worn by the action of the water, is about 70 feet deep, and is crossed by an aqueduct bridge of 88 feet span, lined with cast iron plates, with an eliptical arch of 25 feet rise, resting on stone abutments. Proceeding one mile from this bridge, it passes the two State Prison Farm tunnels; one 416 feet, and the other 375 feet in length; and at a further distance of half a mile, the Holis Brook tunnel is entered, and in passing the valley, the grade line is 35 feet above the stream, and the top filling of the aqueduct, 49 feet. The culvert is 6 feet in diameter, and 131 in length. One mile further it crosses Ryder's Brook, where the foundation wall is 20 feet high, from the bed of the stream to grade, and 34 to the top line of the aqueduct; the culvert is 6 feet by 100. The next object is a culvert or viaduct, erected over the road, with an arch built of stone, of 20 feet span. Proceeding north, the line is conducted through the Austin Farm tunnel, 186 feet long, thence to Mill river there are several valleys, where the depression of the earth varies from 5 to 15 feet below the grade of the aqueduct, and from 20 to 30 feet to the top filling. At Mill River, 13 miles from the dam, the grade is 72 feet above the surface of the river, and the foundation wall, including the aqueduct, reaches to the height of 87 feet. The culvert is 25 feet in diameter, and 172 feet in length. From Mill River the aqueduct passes five depressions, all of which required culverts. Two miles below

Mill River is the White Plains tunnel, 246 feet long; and two and a half miles further, the aqueduct crosses Jewell's Brook, which requires a foundation wall of 50 feet to grade, and 62 to the top of the embankment, constructed with immense labour. The culvert is 6 feet in diameter, and 148 in length. An additional road culvert, 14 feet wide, and 141 long, is also erected here. At a distance of $18\frac{1}{2}$ miles from the Croton Dam, the aqueduct crosses Wiltsey's Brook, 36 feet below the grade, and 49 feet below the top filling, with a culvert 6 feet in diameter, and 137 feet in length. About half a mile beyond this, it passes Dobb's Ferry tunnel, entirely through earth, 262 feet in length. The depression at Storm's Brook, is 29 feet, and 40 feet from the top of the aqueduct; the culvert here is 137 feet in length and 6 in diameter. Between Storm's Brook and Cook's Run, there are several minor depressions. At the latter, the foundation wall is raised about 30 feet above the stream, and the top of the aqueduct is 40 feet; culvert 4 by 32 feet. This terminates the second division.

Entering the third division, the aqueduct proceeds to Dykeman's Brook, 22 miles from the Croton, and 18 from New York. The grade line here is 21 and top of the embankment 35 feet above the surface.

After encountering five unimportant depressions the line reaches the village of Yonkers, 25 miles from the dam, and passes the great tunnel near, and aqueduct over, the Saw Mill river. The tunnel is excavated through earth and rock for a distance of 684 feet. The foundation here is 42 feet above the bottom of the valley, and the top covering of the aqueduct, 56 feet. The culverts are double, and of the largest dimensions; being each 25 feet in diameter, and 90 in length. Another expensive road structure is also constructed here. It has a span of arch of 20 feet, and is $31\frac{1}{2}$ in length, with the proportional height of abutments. After crossing Nodine's Run, the aqueduct passes a considerable elevation by Tibbett's Brook tunnel, 810 feet in length, cut in solid rock. Immediately emerging from the tunnel, the line crosses the valley of Tibbet's Brook, 27 feet below grade, and 40 below the top filling. The culvert is 6 feet in diameter, and 107 in length. Passing many inconsiderable depressions, which each required a culvert, including Aiken's Brook, depressed 28 feet below grade, the aqueduct reaches the crossing of Harlem River, 33 miles from the dam, and 7 from the distributing basin in the city of New York. Here the aqueduct encounters its most formidable impediment. Harlem river, improperly so called, is merely a strait which separates the main land of West Chester county from Manhattan or New York Island. Its length, from the Hudson to the East River is inconsiderable, not exceeding five miles; its depth at the crossing of the aqueduct, 26 feet at ordinary high tide, and its width at the same point, is 620 feet. Owing to the great depression of the stream below the grade line, and the peculiar inclinations of its banks, the length of the aqueduct bridge, will greatly exceed the width of the strait, at its surface (620 feet.) The bridge will be 1,420 feet in length, between the pipe chambers at either end; 18 feet in width, inside of the parapet walls; and 27 feet between the outer edges of the coping; 16 piers, built of stone laid in courses of uniform thickness. Of these, 6 will be in the river, and 10 on the land, (8 of which will be on the West Chester side of the strait.) The river piers will be 20 by 40 feet at base, and 84 feet in height, to the spring of the arch; diminishing as they rise in height. The arches will have a span of 80 feet. The land piers will be proportionably less in size, their height varying according to the slope of the banks, and the span of these arches will be 50 feet each.

The central height of the arches over the stream is to be 100 feet above high water level, in the clear; and the distance from high tide to the top of the parapet walls will be 116 feet. The total elevation of the structure, from its base at the bottom of the strait to the top of the parapet, will be about 138 feet. The piers and abutments will be carried up with pilasters to the top of the parapet, with a projection of two feet beyond the face of the work. Those piers to be erected in the water, will commence with solid rock, upon which the earthy bed of the stream reposes. The estimated cost of this structure is $755,130.

The bridge is intended for the support of iron pipes; and these will be laid down, in the first instance, two of three feet diameter, which it is supposed will be adequate for the supply of water to the city, for many years to come. The work however will be so arranged, as to admit the introduction, at any time hereafter, of two four feet pipes, whose capacity will be equal to that of the grand trunk. The pipes will be protected from the action of the frost, by a covering of earth four feet in depth, well sodded on the surface. The aqueduct will discharge its water into the northern pipe chamber, whence it will pass over the bridge into the southern chamber, where the aqueduct resumes its course towards the city. At the distance of half a mile, the line crosses a ravine of 30 feet to the top line of the embankment; and at a short distance beyond, it enters the Jumel tunnel, 234 feet in length; and 6½ miles from the city. A ravine is passed soon after leaving the tunnel, 25 feet below the grade line; and soon after, another, still more formidable, presents itself; which required a foundation of 30 feet to elevate it to the grade. No impediment of importance occurs until the work reaches Manhattanville, near which occurs a tunnel, 1,215 feet in length, the longest in the whole series. It is denominated the Manhattan Hill tunnel, and is 35 miles from the point of outset at Croton river.

The water will be conducted over the Manhattan valley by means of iron pipes or inverted syphons. The depression of the valley is 105 feet below the grade line, and arrangements of pipe chambers, on each side of the valley, similar to that at Harlem strait, will be adopted here. The pipes are to be laid on a foundation of stone, covered with a course of concrete masonry, six inches thick. After the pipes are laid, concrete is to be worked under them, as a support, 18 inches wide, and 12 high; and the whole is to be protected with a covering of earth, to guard against frost and other injury. The aqueduct having terminated at one pipe chamber, on Manhattan Hills, it re-commences at another on the Asylum Hill; and after proceeding a short distance southward, enters the Asylum Hill tunnel 640 feet in length, which is the last. About three miles from the southern terminus of this herculean work, the aqueduct commences its passage over several streets, the grading of which has a mean depression below that of the aqueduct, of about 40 feet; this vale is to be passed by a bridge of a corresponding height. The line of aqueduct runs 100 feet east of the Ninth avenue; and on the land, extending from one street to the other, a foundation wall is to be built of sufficient width and height to support the aqueduct. Over the carriage way

NEW YORK. 67

and side walks of each street, there will be circular arches turned. Ninety-sixth street, being 100 feet wide, will have two arches of 27 feet span, for the carriage way; and one arch of 14 feet span, on each side, for the side-walks. The other streets, being only 60 feet in width, will each have an arch of 30 feet span for the carriage way, and one on each side, of 10 feet span. The breadth over the arches to be 24 feet.

On the whole line there will be ventilators placed at intervals of one mile apart; and between each, triangular cavities, designed for the erection of additional ventilators, are left, covered with flag stone, and their location indicated by marble slabs. Some of the ventilators can be used as waste weirs and as entrances into the aqueduct.

The next important work is the receiving reservoir, 38 miles by the line of the aqueduct, from its northern terminus. It covers 35 acres of ground, divided into two sections. The north section to have 20 feet of water when full; and the south, 25 feet; the whole reservoir will contain about 160,000,000 of gallons. From this reservoir the water will be conveyed through the Fifth avenue to the distributing basin, of about 5 acres, holding 20,000,000 of gallons, at Murray Hill, in Forty-second street, by means of pipes 30 inches in diameter. From Murray Hill the water will be conveyed to the city by the ordinary distributing pipes.

The difference of level between the basin at Murray Hill and the pool at Croton, is about 46 feet, being a fraction less than 14 inches to the mile.

About 26 miles of the aqueduct are now (April, 1840,) completed, and several other detached sections are nearly so. It must not, however, be inferred that the work still to be done is of but small amount; on the contrary, the most difficult and expensive portions of it remain to be performed. According to the engineers' report, the whole work, with the exception of the bridge over Harlem strait, will be completed and ready for use in the spring of 1842. The completion of the bridge cannot be expected before the close of 1843; and it may and will probably be still further delayed. To diminish this delay, it is proposed to erect a temporary conduit pipe of suitable dimensions, as soon as the coffer dams at Harlem will admit of it, by which means the city may have the benefit of the water, two or three years before a supply could be had by the Harlem aqueduct bridge.

The original estimate of cost of this great work, was $4,718,197; but it will not fall short of $10,000,000;— $3,924,650 08 having been expended at the date of the last report, January 1st, 1840.

In considering this gigantic undertaking, in all its various aspects; its great extent; immense cost; and the length of time which must elapse before it can be made available; the question naturally suggests itself, whether an adequate supply of water, for all the wants of the city, could not have been sooner obtained by other and less expensive means?

The Bronx, it is said by those who have investigated the subject, does not afford such a supply, at all times, as to justify a reliance upon that stream exclusively. Whether any mode of augmenting its volume, by artifical means, has ever engaged the attention of the water commissioners, we know not: but, on glancing at a map, we were forcibly impressed with the apparent fitness of that stream, so far as it extends, to the purposes for which the great aqueduct is designed. The general course of the Bronx, for twenty-five miles, is nearly parallel with, and at a mean distance of only four miles from, the line of the aqueduct: and this course is maintained until it reaches a point within about four miles of the Harlem aqueduct bridge, where it deflects towards the east and passes into Flushing Bay. It rises in the centre of West Chester county and flows in a direction towards the city of New York, with an average descent of about two and a half or three feet to the mile. This is a liberal allowance, as the average grade from the summit level of the New York and Albany Rail-road to tide-water, does not exceed five and a half feet per mile, though the high lands of Sharon are included in the line. Adopting the latter, we have for the sources of the Bronx an altitude above tide-water, of seventy-five feet. Thus far, it is manifest, that the Bronx is, to a large extent, adapted to the purposes of supplying the city with water.

Admitting the objection of an inadequate supply, to be valid, of which we entertain no doubt, the question as to the practicability of turning some of the neighbouring streams into the Bronx, spontaneously presents itself. From the physical structure of the country, it seems probable that such an expedient is highly feasible. The Croton itself could, no doubt, be thus employed; and we were surprised to find that neither Brown, Weston nor Macomb, engineers who were employed to investigate the subject, make any mention of uniting it with the Bronx. Indeed the matter seems to have escaped their attention. The report of Canvass White is the only one among those we have consulted, which suggests such a connection. Although he expresses a doubt as to its practicability, he concludes by saying, "*perhaps a route may be found to connect the Croton with the Bronx or Byram river.*" The surface of the former, at the great reservoir, as we have shown, is $46\frac{9}{100}$ feet above the distributing basin at Murray Hill: and a large part of its course is coincident with that of the Bronx.

The nature of the country and the testimony of travellers, go to prove, that it is much broken by falls and other indictions of considerable elevation, and this is verified by the survey for the New York and Albany Rail-road: and we know that the country situated between the village of Bedford and the Croton valley, has a mean altitude greatly exceeding that of the reservoir. From these considerations, it is obvious that the water of the Croton at its south-eastern bend, is sufficiently elevated to admit its discharge into the ravine of the Bronx, unless insurmountable difficulties of a physical description, of which we are not aware, should interpose to prevent it.

The reader, by turning to a map of New York, will perceive that a small branch of the Croton interlocks with the head streams of the Bronx. This stream flows in a direction from south to north, and with the Bronx valley, forms a natural canal, with one interruption only, extending from the Croton to a point *within four miles of Harlem river*. Such being the hydrography of this region, it only remains to ascertain the nature and elevation of the ridge which divides the waters running into the Croton and Bronx, respectively. Judging from a mere inspection of the map, not the most safe guide, we admit, the extent of tunneling and other excavation and embankment would not exceed seven miles in length, which, added to those necessary to conduct the water from the Bronx to the distributing basin, about eight miles, would make fifteen miles of cutting and embankment, in place of nearly *forty-one miles*, the length of the work now in progress.

The passage of the ridge, just mentioned, would, doubtless, be attended by a cost greatly exceeding that of a similar section, in point of length, of that work : and in addition, the erection of works for the purpose of raising the water into the distributing basin, would be required. These works, however, would supersede the necessity of the elevated and expensive structure across Harlem river, as an aqueduct bridge of an ordinary description would answer every purpose. The difference in cost between the two, would, no doubt, pay the expense of the water-works.

Assuming such a connection to be entirely practicable, an immense amount of labour and of course, expense, would have been avoided by its adoption. Nearly two-thirds of the heavy and expensive excavation and embankment, would have been dispensed with and the aggregate amount of tunneling materially reduced.

Whatever should be the cost of a line, such as we have described, it could not, if found at all practicable, exceed, in any event, that of the works now in course of construction, and would, unquestionably, fall very far short of the actual cost of those works.

RAIL-ROADS.

Long Island Rail-road, commences at the South Ferry in Brooklyn, and proceeding along Atlantic-street, in nearly an east course, passing through Bedford and East New York, it enters the village of Jamaica: thence turning towards the north-east, it advances through Brushville and Clowesville to Hicksville, its present eastern terminus. From Hicksville the line is to be continued to Greenport, about 100 miles from Brooklyn. Two lines have been surveyed for its prolongation, one called the North, the other the South route. The former passes through Woodbury, Huntington and Milltown, where it intersects the South line, and thence proceeds to the point of termination. Length of the finished portion 27 miles.* The

---

* This includes the Brooklyn and Jamaica Rail-road, which has been leased to the Long Island Rail-road Company for 45 years. The capital of the latter is $2,400,000, of which $700,000 have been expended.

Brooklyn section, about 2000 feet in length, ascends at the rate of 200 feet per mile. The grades on the other parts of the road do not exceed 40 feet per mile, and the minimum radius of curvature is 5280 feet. Single track, but graded for two.

Plan of construction.—From Brooklyn to Jamaica, a distance of 12 miles, the rail-way consists of a T rail, supported by cast iron chairs, resting upon stone blocks from the commencement in Brooklyn, to Bedford; and upon cross ties or sleepers of wood from thence to Jamaica. Both the sleepers and the stone blocks are three feet longitudinally of the track from centre to centre. The sleepers are of red cedar sawed 6 inches square and 8 feet long; they as well as the blocks rest upon the bed of the road, from which, loam and clay where they occur, (which is rare, as the soil generally, excepting the top mould, consists of the finest gravel,) have been removed to a depth of 18 inches upon the entire breadth of the road.

The T rail upon this road weighs about 38 lbs. per lineal yard—the chair used to support it upon the sleeper, weighs 15 lbs., and the chair supporting it upon the stone block, weighs 20 lbs.

The opposite stone blocks are connected across the track by an iron tie, consisting of a bar half an inch thick by $2\frac{3}{4}$ inches wide, and 4 feet $8\frac{1}{2}$ inches long; which last is the distance in the clear between the iron rails. In the bottom of the chair used with the stone blocks, is a recess of the width and thickness of this bar, into which it fits, and by the spike which passes through holes corresponding in both chair and bar, they are fastened to the block together.

The top part of the rail rests on the jaws of the chair, in which it is made fast by a double key; the ends of the rails are square, and their lengths fifteen yards.

From Jamaica to Hicksville, a distance of 15 miles, the rail used is of the double T, or H form, 15 feet long with square ends, and weighing $56\frac{1}{2}$ lbs. per yard, resting upon sleepers flatted both sides to a vertical thickness of 5 inches. These sleepers bear upon under longitudinal sills of wood, 3×10 inches. At the joinings of the rails are cast iron chairs, weighing each 8 lbs. with 4 square holes to admit the 4 spikes by which, and their brad heads, the lower web or base of the rail is fastened down upon the chair, and the chair to the sleeper.

The contrivance adopted to counteract an endwise movement of the rail, is a horizontal projection on the top of the chair which fills a square notch cut out of the base of the rail. This provision is upon both sides of the chair, but only at one end of the rail ; that end is in a fixed position, whilst all other parts move and slide upon the supports, in the contractions and expansions of the rail from changes in temperature. At the intermediate bearings the rail is secured by means of brad headed spikes, 4 of which are driven into each sleeper, being one upon each side of the lower web of each rail in each sleeper. The spikes weigh 10 oz. each, and cost 9 cents per lb.

It may here be observed, that the under sills (to support the cross ties) have been adopted between Jamaica and Hicksville, on account of the use of such a support being dictated by the previous experience upon the Boston and Providence Rail-way.

Harlem Rail-road, commences near the City Hall, in New York, and passing along Centre and Broome streets and the Bowery, enters the Fourth Avenue, which it pursues to Harlem Strait, about 8 miles from the City Hall. At its termination a bridge crosses the strait to Morrisania where the New York and Albany Rail-road commences. The road is laid with a double track, and is traversed for nearly three-fourths of its length, by steam power. Owing to the peculiar nature of the ground and the necessity for maintaining a nearly level grade, for a considerable part of the line, long and heavy cuts and embankments were required, which augmented the cost of construction far beyond that of any other similar work in this country. The whole cost of the work, including depots, motive and other power, &c. amounted to $1,100,000 or $137,500 per mile. The receipts for fare, by the company during the year ending December 31st, 1839, were $99,811. Notwithstanding the great number of persons conveyed on this road, about 1,200,000 annually, the directors have not as yet declared a dividend, and up to the 1st of January, 1840, the stockholders had not received a dollar from the work. The tunnel through which the line passes, is the most costly portion, as well as the most attractive feature of the road. Among the thousands who are almost daily conveyed through it, a vast majority is impelled by a desire to examine the " tunnel," which, though excavated at an immense cost, contributes, in no small degree, to increase

NEW YORK. 73

the revenues of the company. The tunnel is cut through solid rock, which chiefly consists of quartz and hornblende of such a compact texture, that masonry is entirely dispensed with, even at the ends. It extends along the Fourth Avenue from 91st to 94th streets, and is 844 feet in length, 24 in width, and 21 in depth from the crown of the arch. The road descends through the tunnel, towards Harlem, at the rate of 25 feet per mile; maximum inclination, 30 feet per mile.

The rail-way consists of the plate rail, $2\frac{1}{4}$ inches wide by $\frac{5}{8}$ thick, laid upon string pieces of wood 7×7, resting upon cross ties of locust or cedar, $3\frac{1}{2}$ feet apart. Within the densely settled part of the city, the plate rail is laid upon stone sills arranged longitudinally, to afford a continuous support to the rail. The opening of this important, though short rail-road, was a desirable event, as it forms the commencement of the great line to Albany. An extension of this road from Centre, through Canal street to the Hudson, is proposed.

NEW YORK AND ALBANY RAIL-ROAD, commences at Morrisania, on the east bank of Harlem river opposite to the termination of the Harlem Rail-road. From thence it proceeds north, through the county of West Chester, nearly equidistant between the Hudson river on the west and Long Island Sound on the east. From the north boundary of West Chester county, the line passes through the eastern part of Putnam and Dutchess counties; the centre of Columbia county, thence to Greenbush, opposite Albany, and also to Troy in Rensselaer county. In its course the line approaches very near the western boundaries of the states of Connecticut and Massachusetts. The road attains its greatest altitude, 769 feet, in the north-east part of Dutchess county. The ascents and descents of the summits, are very gradual, not exceeding 30 feet per mile, the steeper grades being confined to about $\frac{4}{10}$ of the distance. The remainder varies from level to 25 feet per mile. From Harlem to the first summit, 26 miles, a rise of 16 feet per mile. From summit to Croton valley, 12 miles, a fall of 20 feet per mile. Croton to second summit, 54 miles, a rise of $10\frac{6}{12}$ feet per mile. Thence to Greenbush, 47.07 miles, a fall of 16 feet per mile. The line in its northern course, traverses successively, the valleys of Bronx, Croton, Ten-mile rivers; and of Ancram and Cline creeks. The whole distance from the City Hall in New York to Albany is 147.71 miles. The radii of curvature exceed, with two exceptions, 1500 feet. $\frac{3}{10}$ of the road consist of curved, and $\frac{7}{10}$ of straight lines. Estimated expense for a single track, from Harlem river to Greenbush, $2,377,946 74. This is exclusive of land damages, warehouses, locomotives, &c. &c.

The great importance of this thoroughfare will become apparent on glancing at the map. Independently of the immense mineral and agricultural resources, which would be developed in the event of its completion, the amount of travel, especially in the winter season when the navigation of the Hudson is interrupted by ice, would be incalculable. The Housatonic, the Hudson and Berkshire, the Great Western, the Troy, and West Stockbridge and all the western rail-roads, would be auxiliary to it. It would, in short, perform, during one-half the year, all its own duties as well as those of the Hudson, and during the other half, would, if judiciously conducted, participate, largely, in the summer travel, and enter into direct competition with the steam-boats on the Hudson. Its early completion should be steadily aimed at by those who are in charge of this important work.

NEW YORK AND ERIE RAIL-ROAD, commences on the Hudson river, at Tappan in Rockland county, 25 miles above New York, and thence pursues a general north-west course to Goshen in Orange county; from Goshen the line proceeds over the Wallkill by Mount Hope, crosses the Hudson and Delaware Canal, and descends into the valley of the Nevisink, which is traversed for a few miles. Leaving the Nevisink near Monticello, in Sullivan county, it continues towards the north-west, and reaches the outlet of the Popacton branch of the Delaware, and thence along the left bank of the latter to Deposit, where it crosses the Mohawk branch, and proceeds over the dividing ridge between the Delaware and Susquehanna, where it attains an elevation of 1430 feet above tide-water in the Hudson, and arrives at Binghamton in Broome county. Crossing the Chenango Canal at Binghamton, the road is conducted along the right bank of the Susquehanna, through Owego to Tioga point, and thence to Elmira, where it intersects the Chenango Canal. From Elmira the road leaves the Tioga river, joins the feeder of the Chemung Canal and continues parallel with it to Knoxville, where it rejoins the river and enters the Canisteo

NEW YORK. 75

valley: passing up this valley, through Addison and Hornellsville, in Steuben county, the line ascends the summit level, 1780 feet above tide, in Allegany county, where it attains its greatest altitude. It now descends into the valley of the Genesee, and crossing over to that of the Allegany, it enters Olean in Cattaraugus county, where it unites with Genesee Valley Canal. Pursuing the right bank of the Allegany, the road continues till it reaches the Indian village, where it abruptly leaves the river bank, passes into the Connewango valley, thence to an inclined plane, which conducts it into the level below, and finally terminates on the shore of Lake Erie, in Chautauque county. The entire length of this stupendous work is 450 miles. That section of the road from Tappan to Middletown in Orange county, about 50 miles in length, will soon be completed; and that through the counties of Steuben, Chemung and Tioga, has been put under contract. The company was incorporated in 1832, and in 1836, the legislature authorised a loan of the credit of the state to the New York and Erie Rail-road, to the amount of $3,000,000, subject to certain conditions and restrictions. By the terms of the charter, the company is required to complete one-fourth part of the road in ten years; one-half in fifteen, and the whole in twenty years. It may augment its capital to $10,000,000. It may commence the work on receipt of subscriptions to the amount of $1,000,000: is to relinquish the road to the state at cost, with interest at 14 per cent. per annum, should it be required, after the expiration of ten years, and within fifteen years from the completion of the work.

As the grades and curvatures of the road will, no doubt, be materially changed during the progress of the work, we abstain from entering into details on these points at present, reserving for a future edition, a notice of such items as cannot now be satisfactorily described.

HUDSON AND BERKSHIRE RAIL-ROAD, commences at the city of Hudson in Columbia county, curves round towards the north, and gradually inclining east and then south-east, enters the village of Claverack; thence it proceeds in nearly a direct north-eastern course, through Ghent to Chatham corners; thence east into and through Canaan to the state boundary: here the line curves towards the south, descends the valley of the west branch of the Housatonic, and terminates at West Stockbridge in Massachusetts, where it meets the Great Western Rail-road of that state which extends to Worcester. That portion of the road, 2.75 miles long, which lies in Massachusetts, was constructed by a company chartered by Massachusetts, called the West Stockbridge Rail-road Company. The interest of the two companies were sometime since united, and the joint work is now known by the name of the Hudson and Berkshire Rail-road. Length 33 miles ; the grades for 25 miles from Hudson mostly ascend ; three miles of this distance have an inclination of 70, and a quarter of a mile 80 feet per mile. It was opened for public use on the 29th September, 1838.

CATSKILL AND CANAJOHARIE RAIL-ROAD, commences at the town of Catskill, on the west bank of the Hudson, and proceeds in a north-west direction, through the counties of Greene, Albany, Schoharie and Montgomery, and the towns of Athens, Greenville, Rensselaerville, Middleburg, Carlisle and Root, and terminates at Canajoharie, on the Erie Canal. Length 78 miles.

ALBANY AND WEST STOCKBRIDGE RAIL-ROAD, commences at Greenbush, opposite Albany, and proceeds in a general south-east direction, through Rensselaer and Columbia counties, and by Lebanon Springs, to West Stockbridge, in Berkshire county, Massachusetts, with a branch to Pittsfield. Length 41.75 miles; maximum grade 40 feet per mile ; cost, as estimated by the engineer, $647,529 ; or $15,509 per mile, exclusive of depots and apparatus.

RENSSELAER AND SARATOGA RAIL-ROAD, commences at the steam-boat landing, in the city of Troy, and extends through River street, to the foot of Federal street, where it crosses the Hudson to Green Island, at the confluence of the Mohawk and Hudson, by a viaduct 1,512 feet in length, which is also used for ordinary carriages. On leaving the bridge, the line is conducted along the right bank of the Hudson, over the Mohawk, and through the village of Waterford, to Mechanicsville. Here the road deflects towards the north-west, and pursuing that course, crosses the Champlain Canal near the mouth of Anthony's Kill, and thence along its valley to Ballston, where it joins the Saratogo and Schenectady Rail-road.

The superstructure of the Rensselaer and Saratoga Rail-road

is of wood, with flat rails. That of the bridges is also of wood, constructed upon the lattice plan. The piers and abutments are of stone. A large portion of the road is level and the grades low; the maximum being about 30 feet to the mile. The company was chartered in 1832, with a capital of $300,000; which was subsequently increased to $450,000. Length 23.50.

TROY AND WEST STOCKBRIDGE. This work is designed to open a communication with the Great Western Rail-road of Massachusetts. It passes through the towns of Greenbush, Sand Lake, Schodac, Nassau and Chatham, and unites with the Hudson and West Stockbridge Rail-road, at Chatham Corners. Length, from Troy to the junction, about 30 miles.

WEST TROY AND SCHENECTADY RAIL-ROAD. This road will probably pursue the south bank of the Mohawk. It is not yet commenced. Length 15 miles.

WHITEHALL AND SARATOGA RAIL-ROAD, now in progress. About 43 miles in length.

MOHAWK AND HUDSON RAIL-ROAD, extends from Albany to Schenectady. With the exception of an inclined plane in Schenectady, which has an inclination of 1 in 18 and about 6 miles level, the entire road has an ascending grade, varying from 1 in 225 to 1 in 250. The inclined plane in Albany is $\frac{47}{80}$ of a mile in length, and inclines at the rate of 1 in 18. The entire length of the road 15.86 miles. There are six curves; one of 10 chains, on a radius of 700 feet; two of 8 chains each, of 1,100 feet; one of 9 chains, 4,200; one of 6 chains, of 23,000 feet; and one of 10 chains of 4,000.

The excavations are 38 feet wide, and the embankments 26 feet. The deepest excavation is 47, and the highest embankment 44 feet. Greatest altitude, 335 feet above tide water at Albany. The cross sleepers are of wood, 7 inches in diameter, and 8 feet in length. The iron plate is a bar, 9.16 by $2\frac{1}{2}$ inches, with the upper curves rounded to $1\frac{7}{8}$ inches width. The stone blocks are laid three feet apart from centre to centre, on broken stone, and on these the timber rails are placed. The width between the rails is 4 feet 9 inches. The company was chartered in 1826, with a capital of $600,000, or about $38,000 per mile; and the work was commenced in 1830; double track completed in 1833.

78   NEW YORK.

SARATOGA AND SCHENECTADY RAIL-ROAD. This road, in connection with the Mohawk and Hudson Rail-road, forms the common route from Albany to the springs of Ballston and Saratoga. It was commenced in 1831, and was opened for public use, on July 12th, 1832. Length 21.50 miles; single track; cost $297,237. The grades vary from a level to an inclination of 1 in 330. The road bed, 15 feet wide on embankments, and 30 in excavations, including the side ditches.

UTICA AND SCHENECTADY RAIL-ROAD. This road on leaving Schenectady, proceeds in a north-west direction, and passing the village of Scotia, regains the north bank of the Mohawk; thence by the general course of the river, it reaches Amsterdam, in Montgomery county; thence along the left bank of the Mohawk, through Caughnawaga, St. Johnsville, Manheim, Little Falls village, and Herkimer, to the viaduct, by which it crosses the Mohawk, and thence to Utica; where it connects with the Syracuse and Utica Rail-road. Length 77 miles. Commenced in 1834, and completed in 1836, at a cost of $1,540,000, or $20,000 per mile. This road forms the second link in the great chain by rail-road, from Albany to Buffalo and Falls of Niagara. Some of them are now completed, and others in course of execution. Semi-annual dividend in December, 1839, five per cent.

SYRACUSE AND UTICA RAIL-ROAD, is a continuation of the Utica and Schenectady Rail-road. It passes up the south acclivity of the Mohawk, near to, and parallel with, the Erie Canal, which is crossed in entering Rome. Leaving Rome, it re-crosses the Erie Canal, with which it proceeds, by Verona Centre, over the Oneida creek, and through the villages of Canistota, Sullivan, Chittenango, in Madison county, Fayetteville and Orville, in Onondaga, and terminates at Syracuse. Length 53 miles; capital stock $800,000. This is probably the most productive work in the state. According to a statement of the president, the company received for tolls in five months, $117,614; equal to twelve per cent. on its cost; or thirty per cent. per annum.

SYRACUSE AND AUBURN RAIL-ROAD, continues the route to Buffalo. It leaves Syracuse, and on entering Geddes the tracks divide, and after proceeding a few miles they re-unite, and descend the valley of Nine Mile Creek, to Camillus; thence the road turns westward, passes into Elbridge, and across Skaneateles outlet, where it changes its course to south-west, and finally terminates at Auburn, in Cayuga county. Length 26 miles.

AUBURN AND ROCHESTER RAIL-ROAD. A company was incorporated in 1836, with a capital of $2,000,000, for the purpose of constructing a Rail-road from Auburn to Rochester. The whole line is now (April 1840,) under contract. It is grading for a single track from Auburn to Geneva, from Geneva to Canandaigua for a double track, and from Canandaigua to Rochester for a single track; the embankments for the single track are 14 feet in width, and the excavation 26 feet; the whole well ditched. On the portion of double track, the width in excavation is 36 feet, and in the embankment 24 feet on the surface, with proper slopes. The masonry is of undressed stone, except at the Genesee River, where the bridge abutments and piers are dressed.

The grading between Rochester and Canandaigua (29 miles) is about three-fourths finished, and that portion of the road may be put in operation by July next; between Canandaigua and Auburn the contractors are at work only on the heaviest sections, which portion of the road may be ready for use in July, 1841, by which time there will, in all probability, be a continuous line of rail-roads from Boston to Buffalo, of which line the Auburn and Rochester Rail-road will be second to none in importance.

This road will cost less per mile than any of the others east of this, on the same line; the reason of this, to any one acquainted with this country is obvious; on all the other roads from this to Albany they have had more or less difficult and expensive works, such as inclined planes, expensive bridges, river wall, steep side hills, heavy rock cutting and swamps. Now, with the exception of one mile of swamp at the foot of Cayuga lake, there is no work of the above character. Length 80 miles. The total cost of this work, as estimated by the engineer, will be $1,124,710 46. An extension of this road to Lockport whence there is a rail-road to the Falls of Niagara, is proposed.

TONAWANDA RAIL-ROAD. This road extends from Rochester, on the Genesee river in Monroe county, to Attica, in Genesee county. Its course from Rochester to Batavia is west southwest and nearly direct. At Batavia, it assumes a more western direction, which is maintained to the end of the road at Attica. The line traverses the townships of Gates, Chili, and Riga, in Monroe county; and those of Bergen, Byron, Stafford, Batavia and Alexander in Genesee. Length 45 miles.

ATTICA AND BUFFALO RAIL-ROAD. The terminating link in the great chain of rail-roads from Albany to Buffalo. It is 30 miles in length; the greatest inclination, which is at the Buffalo summit, is 35 feet for a distance of two miles; the remainder does not in any case exceed 30 feet per mile. Estimated cost of construction, building, road apparatus, &c. $8,000 per mile.

This, with the road from Rochester to Auburn, when completed, will perfect a continuous line of rail-roads from Albany to Buffalo. The sections which remain unfinished, will, doubtless, soon be executed.

BUFFALO AND NIAGARA FALLS RAIL-ROAD, passes along the bank of the Erie Canal from Buffalo to Black Rock, thence it diverges for a mile or two, and then, resuming its northern direction, descends the valley of Tonawanda, crosses that creek, pursues the right bank of the Niagara strait, opposite to Grand Island, and terminates at the village of Grand Niagara, near the Falls. Length 23 miles; cost $110,000. Company incorporated in 1834.

LOCKPORT AND NIAGARA RAIL-ROAD, extends from the Falls, through the townships of Niagara, Wheatfield and Cambria to Lockport, in Niagara county, a distance of 20 miles. Company incorporated in 1834, with a capital of $175,000.

BUFFALO AND BLACK ROCK RAIL-ROAD. This road is of a peculiar construction, all of wood, except the rails; three miles in length, and cost about $7,500.

ROCHESTER RAIL-ROAD, from the head of navigation in the Genesee river, to Rochester in Monroe county. It descends the right or east bank of the Genesee to the landing at Port Genesee, 255 feet below the Erie Canal at Rochester. Cost $30,000; opened for public use, 1st January, 1833.

ITHACA AND OWEGO RAIL-ROAD, extends from Ithaca at the southern extremity of Cayuga Lake to Owego on the Susquehanna, in Tioga county, where it intersects the line of the proposed New York and Erie Rail-road. It is 29 miles in

NEW YORK.

extent, with two inclined planes which conduct the road from the base up to the summit, 607 feet above the lake. One plane, 1733 feet in length, has a grade of 1 in 28 and the other 1 in 21,. 2225 feet long. Stationary steam power is used on the first, and horse power on the other. Horse power is also used on the other parts of the road. Maximum curvature, 10,000 feet radius—minimum, 700 feet. The curve with which the road enters the town of Owego, has a radius of only a few hundred feet.

BATH RAIL-ROAD, from the town of Bath in Steuben county to Crooked Lake, 5 miles in length.

OGDENSBURG AND CHAMPLAIN RAIL-ROAD. Surveys for this road have been made, the result of which establishes its entire practicability. The length of the most feasible route is about 122.08 miles. The rate of ascent in reaching the summit from Ogdensburg, is 30 feet per mile, and thence to Lake Champlain, 33 feet per mile. Probable cost $1,451,805 05 or about $11,900 per mile.

OSWEGO AND UTICA RAIL-ROAD. To extend between those towns. Length about 70 miles.

PORT KENT AND KEESVILLE RAIL-ROAD, extends from Port Kent in Essex county, to Keesville in Clinton county, four and a half miles in length. Maximum grade 40 feet per mile. Cost $60,000.

Aggregate length of canals in state of N. York, 931.25 miles.
    "   "   rail-roads   "   670.11 miles.

In addition to the works already completed or in progress, the following rail-roads are proposed:

Names.	When incorporated.	Capital authorized.
Adonirac	1839	$100,000
Albion and Tonawanda	1832	250,000
Attica and Sheldon	1836	50,000
Auburn and Canal	1832	150,000
Aurora and Buffalo	1832	300,000
Binghamton and Susquehanna	1833	150,000
Black River Company	1832	900,000

Black River	1836	200,000
Brewertown and Syracuse	1836	80,000
Brooklyn, Ft. H. B. & C. Island	1836	150,000
Buffalo and Batavia	1838	500,000
Buffalo and Erie	1832	650,000
Casadaga and Erie	1836	250,000
Cherry Valley and Susquehanna	1836	500,000
Chemung and Ithaca	1837	200,000
Coxsackie and Schenectady	1837	500,000
Cooperstown and Cherry Valley	1837	150,000
Coldspring	1839	2,500
Coeymans	1836	75,000
Dansville and Rochester	1832	300,000
Delaware	1836	400,000
Dutchess	1836	1,000,000
Erie and Cattaraugus	1837	200,000
Fishhouse and Amsterdam	1832	250,000
Fredonia and Van Buren Harbour	1836	12,000
Genesee and Cattaraugus	1837	400,000
Genesee and Pittsford	1836	150,000
Geneva and Canandaigua	1831	140,000
Goshen and New Jersey	1837	150,000
Great Au Sable	1833	150,000
Gilboa	1839	150,000
Greene	1837	20,000
Herkimer and Trenton	1836	220,000
Honeyoye	1836	250,000
Hudson and Delaware	1830	500,000
Jordan and Skaneateles	1837	20,000
Ithaca and Auburn	1836	500,000
Ithaca and Geneva	1832	800,000
Ithaca and Port Renwick	1834	15,000
Jamesville	1836	25,000
Johnstown	1836	75,000
Kingston and Turnpike	1835	20,000
Lewistown	1836	50,000
Lockport and Batavia	1836	200,000
Lockport and Youngstown	1836	350,000
Malden	1837	350,000

NEW YORK. 83

Manheim and Salisbury	1834	1,500,000
Mayville and Portland	1832	100,000
Medina and Darien	1834	100,000
Medina and Lake Ontario	1836	100,000
Newark	1836	100,000
Oswego and Syracuse	1839	500,000
Oswego and Utica	1836	750,000
Otsego	1832	200,000
Owego and Cortland	1836	500,000
Penfield and Canal	1837	12,000
Rochester and Charlotte	1836	100,000
Rochester and Lockport	1837	400,000
Rome and Port Ontario	1837	350,000
Rutland and Whitehall	1836	100,000
Saratoga and Fort Edward	1832	200,000
Saratoga and Montgomery	1836	150,000
Saratoga and Schuylerville	1832	100,000
Saratoga and Washington	1834	600,000
Schoharie and Otsego	1832	300,000
Scottsville and Onondaigua	1838	100,000
Scottsville and Leroy	1836	200,000
Sharon and Root	1838	70,000
Skaneatelas	1836	25,000
Staten Island	1836	250,000
Syracuse, Cortland and Binghamton	1836	500,000
Syracuse and Onondaigua	1836	75,000
Syracuse and Stone	1836	75,000
Trenton and Sackett's Harbour	1837	600,000
Troy Turnpike and Rail-road	1831	1,000,000
Tyrone and Geneva	1837	500,000
Ulster County	1836	500,000
Unadilla and Schoharie	1836	600,000
Utica and Susquehanna	1832	1,000,000
Warren County	1832	250,000
Warsaw and Leroy	1834	100,000
Warwick,	1837	100,000
Watertown and Rome	1836	1,000,000
Watervliet and Schenectady	1836	500,000
Whitehall and Rutland	1833	158,000

NEW JERSEY.

ALL the canals and rail-roads of this state have been executed by joint stock companies, exclusively. The policy of the state government, with regard to internal improvements, has been one of extreme caution: which, whilst it sanctioned, by legal enactments, the enterprises of individuals and private companies, forbade any direct appropriation of its funds to objects of this description.

Notwithstanding this apparent indifference to these objects, on the part of the state authorities, the work of internal improvement went on, and the state now rejoices in the possession of some of the most important and productive canals and rail-roads in the country; the complete success of which vindicates, triumphantly, the course which prudence and a commendable economy dictated. The transit duties levied by the state upon passengers and merchandise conveyed upon the canals and rail-roads now in operation, *furnish an annual sum sufficient to pay the ordinary expenses of the state government.* These will, no doubt, increase, so as to enable her to become, ultimately, the purchaser of most of them, which, by the terms of the charters, she has the privilege of doing at the expiration of a certain time, on reimbursing the stockholders.

Thus it will be perceived that, although individuals in their corporate capacity, have advanced the necessary funds for the construction of those works, and though New Jersey has not advanced or even loaned a dollar towards it, still the fee is in her, and not in them. They are truly mere lessees for a term of years only, and the state can, and unquestionably will, dissolve all corporations whose works yield a net income, beyond the current interest of the state, whenever that term expires.

The relation that exists between the corporation, in such cases, and the state, is simply that of landlord and tenant, with leave to improve, under limitations and restrictions, dictated by the state and acceded to and ratified by the former. Under these arrangements, the state has abundant reason to be satisfied; she gave nothing and gains every thing; and has thus furnished to her own citizens and the public, a communication as cheap, safe and expeditious as any in the United States, and completed for the country one of the most important links in the chain of communication between the north and south.

From the Delaware and Raritan Canal and the Camden and Amboy Rail-road and its branches, which now form one interest, most of the revenue is derived by the state, but chiefly from the rail-roads. Cost of the works, including land, $6,064,953 42. The company's receipts for the six months ending December 31st, 1839, show a profit of seven per cent. which greatly exceeds that of any former period. With the exception of 1836, there has been from the opening of these works up to the present time, a regular and progressive increase of the net profits, as will appear from the following statement.

An annual Statement, showing the number of passengers and tons of merchandize transported across the State over the Camden and Amboy Rail-road.

Columns A. A. show the *relative* increase and diminution, of the number of passengers and tons of merchandize transported across the state. The year 1833 being estimated at a hundred.

	Number of Passengers.	A.	Weight of Merchandize.	A.
From Jan. 1 to Dec. 31, 1833	109,908	100	6,043	100
" 1834	105,418	95½	8,897	139
" 1835	147,424	134	10,811	178¾
" 1836	163,731	149	12,508	207
" 1837	145,461	132½	10,642	176
" 1838	164,520	149¾	11,765	194½
" 1839	181,479	165	13,520	223¾

NEW JERSEY.

Yearly Statement of Receipts, and Comparative Statement of the same.

No. 1, Date. No. 2, Gross amount of Receipts. No. 3, Comparative statement, showing the relative proportion that the receipts of the different years bear to the receipts of the year 1833. No. 4, Gross expenditures. No. 5, shows the relative proportion that the expenditures bear to the receipts of the year 1833. No. 6, Net gain. No. 7, shows the relative proportion of the net gain to the receipts of the year 1833.

No. 1.	No. 2.	No. 3.	No. 4.	No.5	No. 6.	No.7
From Jan. 1 to Dec.31, 1833	468,142 50	100	287,091 90	$61\frac{1}{2}$	181,050 60	$38\frac{1}{2}$
" 1834	546,993 54	117	313,261 69	67	233,731 87	50
" 1835	679,463 63	146	317,491 76	69	361,971 87	77
" 1836	770,621 28	$165\frac{1}{2}$	363,344 90	78	407,276 38	$87\frac{1}{2}$
" 1837	731,995 24	$156\frac{1}{2}$	359,510 44	77	372,484 80	$79\frac{1}{2}$
" 1838	754,989 89	$161\frac{1}{2}$	355,249 10	76	399,740 79	$85\frac{1}{2}$
" 1839	685,329 76	$146\frac{1}{4}$	258,043 48	55	427,286 28	$91\frac{1}{4}$

From this statement it appears there has been an annual increase of the net profits of the companies of 20 per cent.

Total receipts in 6 years . $4,637,535
Expenditures . . 2,253,993

Net income $2,383,542

This sum exceeds the entire cost of the road, exclusive of the cost of steam-boats, land, and engineering.

CANALS.

DELAWARE AND RARITAN CANAL, commences at Bordentown and proceeds in a north-west direction up the left bank of the Delaware to Trenton ; thence, abruptly turning towards north-east, it ascends the right bank of the Assanpink creek a few miles, then crossing the dividing ridge between the Assanpink and Millstone river, it enters and descends the valley of the latter, along its right or west bank to the Raritan. Here the canal deflects to the south-east, and passing along the right bank of the Raritan, enters the city of New Brunswick where it terminates. This canal, in connexion with the Delaware and Raritan rivers and Staten Island Sound, forms a complete inland water communication between the cities of Philadelphia and New York. The canal intersects the counties of Burlington, Mercer, and Middlesex; and passes through Trenton, near Princeton, Kingston, Griggstown, Millstone and Milton. It was commenced by a joint stock company in 1831, which in the same year, was united with the Camden and Amboy Railroad Company, and under this joint management, the work was continued and finally completed in 1834. The length of the canal is 42 miles; 75 feet wide, and 7 deep; 14 locks, each 100 by 24 feet, and 1 tide lock at New Brunswick; 116 feet of lockage; 17 culverts; 1 aqueduct, and 29 road and farm bridges. Cost, including Delaware feeder, $2,500,000.

The feeder of this canal is also navigable, being from 50 to 60 feet wide at the surface of the water, and six feet deep. It commences at Bull's Island in the Delaware, pursues the left bank of that river, and, on reaching Trenton, curves eastward, and unites with the main trunk. Length 23 miles; grade, descending towards Trenton, 2 inches in a mile; 1 lift and 2 guard locks; 15 culverts and 37 bridges.

MORRIS CANAL, commences at Jersey City, opposite New York, pursues a circuitous route through the Bergen marshes, and crossing the Hackensack and Passaic rivers a short distance above their discharge into Newark Bay, enters the flourishing town of Newark. Here the canal assumes a course nearly north, which it maintains to Patterson, passing the village of Bloomfield. After leaving Patterson, its course is nearly south-west to the Little Falls of Passaic, where it crosses that river, and thence pursues a more western direction, through the little town of Powerville into Rockaway valley; still continuing its western course along the valley of the Rockaway, until it enters the township of Roxbury, it ascends the summit level 2 miles north-west from Drakesville. From the summit at Hopatcong pond, the canal is carried along the left bank of the Musconetcong river, which it crosses one and a half miles south-west from Andover Forge; thence, assuming a south-west direction, it passes near the village of Hacketstown, Beatystown, Anderson,

Mansfield, Broadway, and New Village, and terminates on the Delaware at Phillipsburg, opposite Easton. General course from New York to Easton, west: length, 101.75 miles; ascent, 915, descent, 759 feet; total rise and fall, 1674 feet; overcome by locks and inclined planes. The latter consist of apparatus for the purpose of conveying the boats from one level of the canal to another. There is a lock at each end of the plane; one at the foot, in which the boat is adjusted for its ascent, and another at the top to elevate it to the level above; when adjusted, the whole is drawn up by means of appropriate machinery, which is also used for regulating the downward passage of the boats. By means of this ingenious contrivance, which supersedes the necessity for water, as in ordinary locks, the boats are conveyed safely and expeditiously, up or down the several ascents and descents of the line. Elevation of Easton 161, and summit level, 915 feet above the Atlantic; 32 feet wide at top; 20 at bottom; 4 feet deep. Rise and fall, 1674 feet, of which 235 feet are overcome by 24 locks, and 1439 feet by 22 inclined planes, the average inclination of which is about 2 in 21; 4 guard locks; 5 dams; 30 culverts; 12 aqueducts, including one of stone at the Little Falls of Passaic, with a single arch of 80 feet span, and another of wood over the Pompton river, 236 feet in length, supported by nine stone piers; 200 bridges. Cost $3,100,000.

The company under whose direction this important work was executed, was incorporated with banking privileges. Their bank is located in the city of New York.

SALEM CANAL, extends from Salem creek to the Delaware; and is designed to shorten the distance from the upper parts of Salem county to Philadelphia. Length, 4 miles.

RAIL-ROADS.

CAMDEN AND AMBOY RAIL-ROAD, commences at Camden, opposite Philadelphia; thence, crossing Cooper's creek, it follows, generally, the east bank of the Delaware. Six miles from Camden the road crosses Pensaukin creek; and six miles farther it crosses Rancocus by a substantial viaduct. The road hence to Burlington, (six miles) is perfectly straight, and from Burlington to Bordentown, a further distance of ten miles, it is nearly so.

The general direction of the road from Bordentown to South Amboy, its point of termination, is nearly north-east. At a distance of 2½ miles from Bordentown, the road passes Crosswick's creek; 7 miles farther it enters the village of Centreville; thence to Hightstown 4 miles; thence, by nearly a direct course, to Spottswood, 11 miles. About one mile beyond the latter place, the road recrosses the turnpike, and passes Herbertsville two miles from Spottswood. At a further distance of 7 miles this important work terminates at the long wharf in South Amboy. Length from Camden to Bordentown, 27.50 miles, and thence to South Amboy, 33.50 miles. Total length from Camden to South Amboy, 61 miles; commenced in 1830 and completed in 1837; cost $1,238,000. The curves, with a few exceptions, have radii of 1800 feet and upwards. A considerable portion of the road is level, and the inclination on the other parts do not exceed 20 feet to the mile, except at Crosswick's creek, South river and Amboy. At the latter place the grade is 45 feet to the mile, for a short distance.

Plan of construction.—The rail is of the H pattern, in bars of 16 feet in length, with square ends, weighing 41 lbs. per yard. In the track extending from Bordentown to Amboy, the rail is supported on stone blocks 18 inches square by 12 inches deep, laid 3 2-10 feet apart from centre to centre, upon stone broken to pass through a two-inch ring, placed in a continued longitudinal trench 3 feet wide and 2 feet deep, under each rail. The broken stone were well compacted by rollers. At the joinings, as well as at the middle of the rails, are cast or wrought iron plates, with holes in them to admit the spikes, the brad heads of which catch over the base of the rail on each side, out of which, at both the middle and the ends, a notch is cut to admit the shank of the spike, with a view to prevent the endwise movement of the rail. The rails are attached to each other, at their joinings, by a wrought iron plate about 4 inches long, which fits into the hollow of the rail on one side of it, between the upper and lower webs, against the vertical stem, and is fastened to each rail by a rivet having a horizontal position, and passing through the stem of the rail by means of a hole made oblong in one end of the rail, to allow of their contraction and expansion.

Upon the top of each stone block is placed a piece of plank from 1 to 2 inches thick, intervening between the block and the base of the rail, and the joinings of the rails are so ordered that the end of one comes opposite the middle of another, or nearly so. The two lines of rails are tied across the track, at distances of 8 feet, by iron rods passing under the bottom of the rails, and bent over the outside of the lower webs.

The part of the railway between Camden and Bordentown, subsequently made, is laid with wooden cross ties or sleepers, instead of stone blocks, and at the same distance apart, namely, 3 2-10 feet. These sleepers rest upon plank 3 inches thick by about 12 inches wide. The clay and other material retentive of water, and therefore likely to be affected by frost, is removed altogether from the bed of this part of the road, and sand and gravel substituted, (where the latter do not form the graded surface previously,) for a depth of 18 inches.

A portion of the rails used upon this road were imported from England, modified somewhat from the form at first adopted, by making the inner part of the upper web thicker, so as to withstand more effectually the action of the wheels, and to diminish the wear of the flanges, by increasing the breadth of the surface against which they rub.

By the terms of the charter which prohibits the construction of any other road within five miles of the one now in use, this company enjoys a complete monopoly in the conveyance of passengers and merchandize between Philadelphia and New York. The number of the former is immense, averaging during the travelling season about one thousand daily.

TRENTON BRANCH of the Camden and Amboy Rail-road, extends from the main line near Bordentown to Trenton. Length, about 8 miles.

JOBSTOWN BRANCH of the Camden and Amboy Rail-road, from Craft's creek to Jobstown. Length 13 miles.

PATERSON AND HUDSON RAIL-ROAD, commences on the New Jersey Rail-road, about two miles from Jersey City, and thence proceeds in a north-west direction, over the Hackensack and Passaic rivers, to Paterson in Passaic county, a distance of 16.30 miles. Near its point of outset the road passes Bergen ridge

NEW JERSEY. 91

by an excavation of nearly 50 feet in depth, some of it through solid rock. A large portion of the route consists of marsh embankment, but little above the ordinary high tide of the adjoining bay.

The rivers are crossed by substantial bridges, one 1700 feet long, so constructed as to admit the passage of vessels. Company incorporated in 1831, with a capital of $250,000, which may be increased to $500,000.

CAMDEN AND WOODBURY RAIL-ROAD, extends from Camden, on the Delaware opposite Philadelphia, to the town of Woodbury in Gloucester county, 9 miles in length.

NEW JERSEY RAIL-ROAD, extends from the ferry dock in Jersey City, opposite New York, to New Brunswick. Its course to Newark is west-north-west, and thence to New Brunswick, south-west, through the counties of Essex and Middlesex. At a distance of two miles from Jersey City, the road is intersected by the Paterson and Hudson Rail-road; at 10 miles it passes through Newark; at 16 miles, it enters Elizabethtown; at 20, it enters Rahway, and at 34, it crosses the Raritan and terminates at New Brunswick, where it meets the Trenton and New Brunswick Rail-road, a part of the great mail route, towards the south. The company was incorporated in 1832. Cost of road and appendages, $2,000,000. The least radius of curvature, in Newark only, is 400 feet; with this exception, there is none less than 2000 feet. The highest grade is 26 feet per mile. The deep cut through Bergen Hill and the viaduct over the Raritan, deserve especial notice.

Plan of construction.—From the wharf at Jersey City to Newark, two tracks are laid, upon both of which, as far as the branching off of the Patterson Rail-road, at a distance of two miles, and upon one of which from thence to Newark, the plate rail was originally used, laid upon a wooden structure, consisting of string pieces, resting upon notched cross ties. This part of the track has been relaid in a more substantial manner, after the plan next to be described.

Upon one of the two tracks, from Bergen Hill to Newark, 6 miles, and upon the single track from Newark to New Brunswick, 22 miles, the rail is of the T form, weighing about 37 lbs. per yard; each bar being 18 feet long, with square ends. The rail rests by its top part, or two upper flanges, upon the top of the two checks of the chair, whilst its vertical stem descends into the jaws of the chair where it is tightened by means of a single key, fitting into a grooved notch, partly in the chair, and partly in the rail.

The rail is supported at equal distances of three feet from centre to centre, upon the cast iron chair of 15 lbs. each, resting on cross sleepers of wood, and fastened to them by means of two spikes, each 3-8 square and 5 inches long. These cross ties are of red cedar and chesnut; the cedar being sawed 4×5 inches and 7 feet long; the chesnut ties are of the same length, but of somewhat larger section, and were procured in the vicinity at twenty-five cents each. These cross ties are supported upon under-sills of chesnut 5×7 inches at the smaller end, and not less than 18 feet long, laid longitudinally of the track, under each rail. The rail-way is new and in very fine adjustment, and the machinery works well upon it.

Besides two wooden viaducts, the one over the Passaic, and the other over the Hackensack, there are two extraordinary works upon this road that merit a passing notice, namely, the cut through Bergen Hill, and the viaduct over the Raritan at New Brunswick.

Deep Cut through Bergen Hill.—The total length of this cut is about one mile, and the greatest depth 50 feet, of which 35 is in rock, covered by 15 feet of earth. Whole quantity of excavation, five hundred thousand cubic yards, of which two hundred thousand are of hard, silicious, hornblende rock. The average of excavating and removing the rock, is $1 70 per cubic yard; whilst the excavation of the earth near the northern end of the cut cost 15 cents, and at the southern end 10 cents per cubic yard. The breadth of the cut at bottom is 28 feet, and the established grade for the road 26 feet per mile. The cost of excavating a red shale, in a thorough cut of 15 feet in depth, near New Brunswick, was from 40 to 60 cents per cubic yard.

The Raritan viaduct is on Col. Long's plan, and is 1700 feet in length, in spans from 112 to 145 feet reach. Depth of truss 22 feet; width between hand rails on top 31 feet; piers, 7 in number, which with the two abutments, are faced with sienitic granite, from Connecticut, and the filling is of the blue and red shales of the vicinity. The structure is of two stories; the lower floor resting upon the bottom of the trusses, of which there are three, supports a double road way to accommodate common carriages. The railway reposes on the top of the trusses, supported by joist bearers 4 feet apart. The chairs, holding the rails, rest on string pieces, 4 inches thick by 11 wide, pinned down to the upper floor, which latter performs the office of a roof. The braces of the truss framing abut upon pieces of thin sheet iron, introduced into the joints. At a depth of 9 feet from the tops of the piers and abutments, there is an offset of 9 inches, upon which are footed the shore braces that assist in supporting the trusses.

There are four distinct sliding draws, two in each story. The rail-road draws move back into the place of a section which slides sideways, out of the way, while the common road draws roll on, opening over the part of the bridge back of them; a moveable platform connecting the draws with the floor of the bridge, being raised up from it by means of lever beams, when the draw is about to be opened for the passage of vessels. The spans of the draws are each 30 feet, and those in the rail-road cost from $3,000 to $4,000 each.

TRENTON AND NEW BRUNSWICK RAIL-ROAD. This road, which completes the line between Philadelphia and New York, on leaving Trenton, passes along the immediate south-east bank of the Delaware and Raritan Canal. At Kingston it diverges from the canal and pursues the valley of Heathcote's brook a distance of about four miles, to Long Bridge farm, and thence into and along the valley of Lawrence brook to Dean's mill pond; it here ascends and crosses the table land to New Brunswick. The execution of this work was commenced on the 6th of June, 1838, and completed on the 1st of December of the same year. The superstructure is composed of T rail, 16 feet long, resting on 9 cross ties, 8 of which are oak or chesnut, the joint tie being locust, resting on two stone blocks. The rails are united by a cast iron chair. Length, 27 miles.

MORRIS AND ESSEX RAIL-ROAD, branching off from the New Jersey Rail-road in the city of Newark, crosses the Mount Pleasant turnpike, near Orange, and turning towards the south-west, proceeds to South Orange, along the valley of the east branch of Rahway river, to the village of Jefferson in Essex county. Thence by a circuitous western course, to Chatham, and thence through Union Hill and Madison, to Morristown. Length, 22 miles.

MINE BROOK RAIL-ROAD, is to commence at Newark or Elizabethtown, pass through the towns of Springfield, Basking Ridge, Germantown, Clinton, &c. and the counties of Somerset, Huntingdon and Warren, and terminate on the Delaware, opposite Easton, Pa., 75 miles in length; maximum grade 48 feet per mile; greatest curvature, 1000 feet radius.

WHIPPANY RAIL-ROAD, extends from the Morris and Essex Rail-road at Madison, through Columbia, to Whippany. Length about 10 miles.

ELIZABETHPORT AND SOMERVILLE RAIL-ROAD, extends from Elizabethport in Essex, to Somerville in Somerset county. Length 25 miles. It is now completed and in use from Elizabethport on Staten Island Sound, to Plainfield, a distance of 15 miles.

Aggregate length of canals in New Jersey, 170.75 miles.
" " rail-roads " 215.30 "

PENNSYLVANIA.

It is now about eighty years since the first movements were made to introduce a system of internal improvements into the then province of Pennsylvania. The friends of the system were indefatigable in their efforts to promote its commencement, in which they were seconded by the public authorities of the time. Essays showing the utility of internal navigation were written and extensively circulated, by which the people were stimulated to active exertions, in order to secure the accomplishment of this important object.

Having by these means enlisted the zealous co-operation of some of the most influential and wealthy inhabitants of the province on behalf of the proposed measure, application was made to the provincial legislature, for authority to open a water communication between the Schuylkill and Susquehanna rivers, and in the year 1762, a survey with a view to this object, was effected, by which its practicability was satisfactorily demonstrated. In 1791, the "Schuylkill and Susquehanna Canal" was commenced; and in 1794, one of the western sections, four miles in length, was completed and opened for navigation. From this period the further prosecution of the work was suspended; and it was not again resumed until the year 1816, when a newly organized company assumed its management, under whose direction the canal was completed and opened for use in 1824. This is, briefly, the history of the "*Union Canal*" now so called. Other works, less important in character and extent, had been executed in some parts of the province long prior to the above date. They are, however, merely adverted to now in justification of the claims of Pennsylvania, to credit, as the leader in the march of internal improvement in our country. By a reference to early enactments, especially those embraced in the period from 1780 to 1800, it will appear that the legislature was not wholly indifferent to the promotion of internal improvement, or insensible to its importance. A navigable communication between the eastern waters and those of the Ohio, early attracted the attention of the public authorities; surveys for this object, were made by several distinguished engineers, assisted by David Rittenhouse, in his capacity of astronomer, who reported that "the whole distance of a navigation by water between Philadelphia and Pittsburg, would be 426 miles, in which there would not be any interruption but one portage of 18 miles at Conemaugh." The route as proposed by Rittenhouse and his colleagues, coincides, very nearly, with the line of the Union Canal, the central and western divisions of the Pennsylvania Canal, now constructed. In 1792 a company was incorporated to construct a canal between the Delaware, at Philadelphia, and the Schuylkill, at Norristown. This work was subsequently commenced, and considerable progress made in its construction, when, for want of funds and other causes, its further prosecution was suspended for the time, and ultimately abandoned altogether. A part of this line now forms the bed of the Columbia Rail-road, from Fair-mount to a point near Peters's Island. The failure of this enterprise, and the suspension of others of a like description, seem to have paralized the energies of the friends of internal improvement. With the exception of the Schuylkill Navigation and some minor works, little or nothing was done by them towards the introduction of a general system of improvement, until aroused by the example of their neighbours of New York, whose successful achievements now began to attract universal attention, they resumed their efforts, and succeeded in arresting the attention of the government. The legislature of Pennsylvania, actuated by a due sense of the importance of the subject, authorized the immediate commencement of several extensive canals and rail-roads; and on the 4th of July, 1826, was commenced that great system of internal improvement, which for extent, magnitude and utility, stands unrivalled in modern times. Though the canals and rail-roads are now in a condition to accommodate the present trade, there are yet some sections under contract, and in progress, which must be completed in order to perfect the system.

The whole of these public works are so located as to penetrate those sections of the state, which, from their known fertility and mineral resources, afforded the greatest promise of a successful commerce with the great eastern and western emporia of the state; and more, perhaps, than any other sections, required those facilities which would most effectually develop their resourses.

The entire expense to which the state will have been subjected, when the various lines of improvement are completed, will probably not be less than $25,000,000. In addition to the state works, there are distributed throughout the commonwealth, completed or in progress, canals and rail-roads, the aggregate cost of which, when finished, will not fall short of $25,000,000.

CANALS.

Central Division of the Pennsylvania Canal. This canal, with the Columbia and Portage Rail-road, and the western division of the Pennsylvania Canal, forms the great chain of communication between the Delaware and Ohio rivers. It commences at Columbia, on the western terminus of the Columbia and Philadelphia Rail-road, follows the east bank of the Susquehanna, and passes through the villages of Maytown, Bainbridge and Falmouth; intersects the Union Canal at Middletown, where there is a side lock of three feet, connecting this division with the Union Canal, and others, which connect it with the Susquehanna river. From Middletown, after uniting with the Susquehanna by a series of locks, it continues along the east bank of that river, through Highspiretown and Harrisburg to Duncan's Island, where it is intersected by the Susquehanna division of the Pennsylvania Canal. At the head of this island the canal crosses the Susquehanna and enters the valley of the Juniata, which it pursues mostly along its north or left side, and passing Millerstown, Mexico, Mifflintown, Lewistown, Huntingdon and Petersburg; and through the counties of Lancaster, Dauphin, Perry, Juniata and Huntingdon, terminates at Hollidaysburg, where it meets the Portage Rail-road across the Allegany mountain.

Length 172 miles; course W. N. W.; total lockage, from the basin at Columbia, to that at Hollidaysburg 670.53 feet; 40 feet wide at top, 28 at bottom, 4 feet deep; 18 dams; 33 aqueducts; 108 locks, exclusive of 2 guard locks, and outlet locks at Columbia; those between Columbia and Duncan's Island are each 90 by 17 feet; and those from Duncan's Island to Hollidaysburg, 90 by 15 feet in the chamber; the latter are built on the composite plan. 15.83 miles of this canal consists of slack water navigation.

RAYSTOWN FEEDER, at the mouth of the Raystown branch of the Juniata. Length 1 mile.

WESTERN DIVISION OF THE PENNSYLVANIA CANAL. This link in the grand chain which stretches from Philadelphia to Pittsburg, traverses the valleys of the Conemaugh, Kiskiminetas and Allegany rivers, to its termination at Pittsburg. After leaving Johnstown, it passes the towns of Fairfield, Lockport, Blairsville and Saltzburg in Indiana county, Warren and Leechburg in Armstrong, and, crossing the Allegany above the mouth of the Kiskiminetas, enters Butler county, and thence along the right bank of the Allegany (which is crossed by a splendid aqueduct) enters and passes through the city of Pittsburg, and terminates on the Monongahela river. Length 104.25 miles; 40 feet wide at top, 28 at bottom, 4 feet deep; lockage 471 feet; 66 locks (exclusive of four on a branch canal to the Allegany) 90 by 15 feet within the chamber; total lockage 470 feet; 10 dams; 21¼ miles of the above canal consist of slackwater navigation. The average fall from Johnstown to Blairsville, 30 miles, is about 8 feet per mile. Between the latter and Pittsburg it is 3 feet per mile; 2 tunnels; 16 aqueducts; 64 culverts; 39 waste wears; 152 bridges. Cost, $3,000,000.

This work completes the route by rail-roads and canals to Pittsburg. It is one of the great thoroughfares from Philadelphia to the west. The entire distance from Philadelphia by the canal and rail-road is 394.54 miles. A navigable feeder extending from Kittaning to this division and commenced in 1838, was suspended in 1839, as no appropriation had been made by the legislature, for the prosecution of the work. Length 14 miles. Estimated cost $662,603.

JOHNSTOWN FEEDER, at the eastern terminus of the western division. Length 1.50 miles.

ALLEGANY BRANCH, from Alleganytown to the Western division. Length 0.75 miles.

PENNSYLVANIA. 99

SUSQUEHANNA DIVISION OF THE PENNSYLVANIA CANAL, commences at the outlet lock on Duncan's Island, where it joins the Central Division, crosses the north outlet of the Juniata, and enters Buffalo township, in Perry county; thence it pursues nearly a due north course through Perry and Mifflin counties, along the right bank of the Susquehanna, in Union county, to the town of Northumberland, where it intersects the canals which extend along the north and west branches of the Susquehanna. Length 39 miles; cost of construction $1,039,256; 12 locks; and 86.50 feet of lockage.

WEST BRANCH DIVISION, leaves the Susquehanna Division at Northumberland, and passes along the left bank of the west branch of the Susquehanna, through Northumberland and Lycoming counties, to Farrandsville, in Clinton county. Length, including several sections of pool navigation, 73 miles; lockage 138.50 feet; 19 guard and lift locks; 8 dams, varying from 4 to 10 feet in height; 4 aqueducts; cost $927,388.

BALD EAGLE SIDE CUT, extends from the pool at Dunnstown Dam, on the West Branch Division to Bald Eagle Creek. Length 3.62 miles.

LEWISBURG SIDE CUT, extends from Lewisburg, in Union county, to the West Branch Division. Length 0.63 miles.

TANGASCOOTAC EXTENSION, from Dunnstown to the mouth of the Tangascootac, a distance of 7.50 miles; nearly completed.

SINNEMAHONING EXTENSION. This work had progressed to some extent, when, in July, 1839, active operations were suspended for want of funds. It is 33 miles in length, and extends from the mouth of Tangascootac to that of the Sinnemahoning. The line, as located, will require 2 dams; 2 guard, and 17 lift locks; 5 aqueducts; 19 culverts; 10 waste weirs; and 30 bridges; with a lockage of 150 feet; estimated cost $1,388,099 15.

NORTH BRANCH DIVISION, commences at the basin which unites the Susquehanna and West Branch Divisions at Northumberland. This canal pursues a north-eastern course, through Northumberland, Columbia and Luzerne counties; and by the towns of Danville, Bloomsburg, Berwick, &c., and terminates at Lackawana, in Wyoming valley. Length, including pool navigation, 72.50 miles; 40 feet wide at top, 28 at bottom, 4 feet deep; 7 lift and one guard locks, each

PENNSYLVANIA.

17 by 90 feet within the chambers; rise 68.89 feet; cost $1,096,178.

NORTH BRANCH EXTENSION, in progress, commences at the mouth of Lackawana creek, and terminates at the village of Athens, in Bradford county. Length 90 miles, with 189.50 feet of lockage; 23 locks; 6 guard locks; 9 aqueducts; 3 dams; 23 waste weirs; 26 culverts; and 100 bridges. The design of this work is to effect a communication by means of the Chenango Canal of N. York and the Susquehanna, between the improvements of Pennsylvania, and the Erie Canal of New York, and thus facilitate the exchange of the various products of the respective states. Estimated cost of the North Branch Extension $3,528,302 20.

WISCONISCO CANAL, in progress, extends from Wisconisco creek, at the western terminus of the Lykins Valley Rail-road to the pool of Clark's Ferry dam, at Duncan's Island. Length 12.25 miles. It has 1 guard and 6 lift locks; 3 aqueducts; 1 dam; 2 culverts; 5 waste weirs; and 18 bridges; descent 35 feet; estimated cost $376,195 43.

LACKAWANA FEEDER, at the northern terminus of the North Branch Division. Length 0.25.

DELAWARE DIVISION OF THE PENNSYLVANIA CANAL. Commences at Bristol, in Bucks county, 18 miles above Philadelphia.

On leaving the basin at Bristol, the canal passes in nearly a direct line to Morrisville, opposite Trenton, leaving Tullytown on the right, and Tyburn on the left. On leaving Morrisville it pursues a course nearly at right angles with the section from Bristol to Morrisville. Following this direction, sometimes approaching within a short distance of, and then receding from the Delaware, it successively passes the towns of Yardleyville, Taylorsville, Brownsburg, New Hope, Lumberville, Smithville and Monroe, and terminates at Easton, where it unites with the Lehigh Company's works. Length 59.75 miles; 40 feet wide at the water line, and 5 feet deep; rise 164 feet, overcome by 23 lift locks, 90 by 11 feet, varying in height from 6 to 10 feet; 9 aqueducts; 20 culverts; 125 bridges; 2 guard locks; 1 outlet and 1 tide lock; completed, October, 1830; cost $1,275,715.

BEAVER DIVISION OF THE PENNSYLVANIA CANAL. This canal is merely preparatory to a more extensive line of canals designed to connect Conneaut lake, in Crawford county, with Lake Erie, and ultimately with the Ohio river at Pittsburg. Another section of this work is just completed; it extends from the town of Beaver, on the Ohio, to the head of slack water navigation on the Shenango, 6 miles above New Castle, and is 30.75 miles in length; 17 locks; lockage 132 feet. The section of the Beaver Division, from the Ohio to the confluence of Big Beaver and Mahoning, about three miles below Newcastle, forms a part of the communication between the canals of Ohio and Pennsylvania. The surface of low water in the Ohio river at the mouth of Big Beaver is 419.50 feet below the surface of Conneaut lake, and 90.50 above that of Lake Erie.

FRENCH CREEK FEEDER, extends from the head of navigation in the pool of Bemus Dam, three miles above Meadville, to the junction with the Erie extension; including Conneaut Lake. Length 27 miles.

FRANKLIN LINE, extends from the feeder aqueduct over French creek, seven miles below Meadville, to the town of Franklin, on the Allegany river. The surface of the water in the aqueduct will be level with the Conneaut Reservoir when full, and 510 feet above Lake Erie. Length 22.25 miles; lockage 128.50 feet.

ERIE EXTENSION. This work, now in progress, commences at the head of the pool, on the Shenango, six miles above New Castle, Mercer county, and thence proceeds towards the north along the valley of the Shenango, and through Crawford and Erie counties, to Presque Isle, at the town of Erie. It is 105.50 miles in length, and is divided into two lines, viz., the Shenango line, extending to the summit at Conneaut lake, 60 miles; and the Conneaut line, thence to lake Erie, 45.50 miles. The ascent from the New Castle pool to the summit, when the reservoir is full, is 287.50 feet; and the descent thence to the surface of Lake Erie, is 510 feet; making the entire lockage 797.50 feet. A section of 43 miles, extending from the northern terminus of the Newcastle pool to Greenville, in Mercer county, is nearly completed, and will be opened for the public, in the spring of 1840. There are on the Shenango line 44 locks; 35 of stone, and 9 of the composite order, (stone walls faced with plank); 5 dams; 3 aqueducts; 21 road, 46 farm, and 11 towing-path bridges; and 24 waste wiers; estimated cost $1,658,679 06.

PENNSYLVANIA.

THE CONNEAUT LINE. Eleven and a half miles, 13 locks, 6 culverts, and 2 bridges of this line are progressing towards completion. The remainder of the work was put under contract in 1839. It is divided into 61 sections, comprising 21 composite locks; 2 aqueducts; 5 culverts; 50 waste weirs; 41 road, and 40 farm bridges. Estimated cost of the Conneaut line $1,612,515 45; or $3,271,194 51 for the whole work, extending from the pool above Newcastle to Lake Erie.

RECAPITULATION.

Pennsylvania Canal,	Central	Division	. .	172.00
"	"	Western	" . .	104.25
"	"	Susquehanna	" . .	39.00
"	"	W. Branch	" . .	73.00
"	"	N. Branch	" . .	72.50
"	"	Delaware	" . .	59.75
"	"	Pittsburg and Erie*	" . .	80.00
"	"	West Branch Feeder	.	4.25
"	"	North Branch Feeder	.	0.25
"	"	West. Division Feeder	.	2.25
"	"	Cent. Division Feeder	.	1.00
		Total Canals		608.25
Columbia Rail-road		. . . - .	.	81.60
Allegany Portage		. . . .	.	36.69
		Total Rail-roads		118.29
		Total Canals		608.25
	Grand total of Pennsylvania state works			726.54

All the above feeders are navigable. There are others not navigable. That at Hollidaysburg is 3 miles, and that at Swatara is 2 miles in length.

---

* This includes the French Creek Feeder, and all the other improvements in that section of the state.

Canals and Rail Roads constructed by Joint Stock Companies.

CANALS.

SCHUYLKILL NAVIGATION. This work extends from the dam at Fairmount, near Philadelphia, to Port Carbon in Schuylkill county. It consists of a succession of canals and pools. The pool above Fairmount dam is entered by a short canal on the west side of the Schuylkill. This pool extends about six miles to Flat Rock. About $1\frac{1}{2}$ miles from Flat Rock dam, the Manayunk Canal leaves the last mentioned pool and rejoins the stream a short distance above the dam, and thus enters the second pool. This extends 4 miles to an inconsiderable canal, which connects it with the pool above. Three miles farther, another small canal conducts into the pool above Norristown, in Montgomery county; thence the stream is ascended by several short canals and pools, to the commencement of the "Oaks Canal," $3\frac{1}{4}$ miles in length. The Oaks Canal commences half a mile above the outlet of Perkiomen creek, and extends along the north or left bank of the Schuylkill, to a dam about one mile above Phenixville, in Chester county, where it enters the river. The pool formed by the dam just mentioned, extends to the outlet of the Vincent Canal, nearly five miles in length. About 1 mile above the termination of the Vincent Canal, commences the Girard Canal, one of the most extensive in the series. It is 22 miles in length, and extends along the right bank of the Schuylkill, from Pigeon creek five miles below Reading in Berks county. In the space between that point and Reading there are two dams and a small canal, which is connected with that passing through Reading, where the Union Canal intersects the Schuylkill Company's works. The latter canal enters the Schuylkill a short distance below Reading, and is on the left bank of the stream. With the exception of the Hamburg canal, ten miles in length, and another of three miles, the distance from Reading to Hamburg is traversed by a succession of short canals, mostly on the left bank of the Schuylkill; this is also the case from Hamburg to Port Carbon, where the navigation ceases.

This work opens a direct communication between Philadelphia and the anthracite coal region, in Schuylkill county, whence immense quantities of it are transported to Philadelphia and other cities of the Union. Length of canals 58, and of pools 50 miles. Entire length from Fair Mount to Port Carbon, 108 miles. Canals 36 feet wide at top, 22 at bottom, 3 feet 6 inches deep; 129 locks, each 80 by 17 feet; 34 dams; 1 tunnel, 385 feet in length; rise 610 feet; cost $2,500,176. Commenced in 1815; completed in 1826.

FAIRMOUNT WATER WORKS. The hydraulic works by which the city of Philadelphia and the adjoining districts are supplied with water, are situated on the east bank of the Schuylkill, two miles north-west from the city. They occupy an area of 30 acres, which extends from the Schuylkill on the west, to Fairmount street on the east, and from Callowhill and Biddle street on the south to Coates street and the Columbia Rail-road on the north. The greater part of this area consists of the "mount," an oval shaped eminence, about one hundred feet in height, with sides more or less inclined, according to the nature of the formation and the uses to which they are applied.

On the top of the hill, at an elevation of one hundred feet above mid-tide in the Schuylkill, and about 56 feet above the highest ground in the city, there are four reservoirs, whose aggregate capacity is about twenty-two millions of gallons. One of these is divided into three sections, for the purpose of filtration. They are inclosed by a substantial pale fence, which while it serves to protect, does not obstruct the view of the reservoirs. The whole is surrounded by a gravel footway, extending along the entire brow of the hill, which is attained by a flight of steps on the west, and several inclined planes, of easy ascent, from the east.

Fairmount originally extended to, and formed the immediate bank of the Schuylkill, and the entire site of the various structures, and the beautiful embellishments which now adorn the place, and render it an object of peculiar attraction, is the result of expensive and laborious excavation into solid gneiss rock. It was commenced in 1819, and continued with occasional intermissions from that time down to the present day. The requisite power for propelling the machinery, is obtained by means of a pool formed by a dam, erected across the Schuylkill, which backs the water for several miles, and thus serves the double purpose of improving the navigation of the river, and giving motion to the wheels and forcing pumps by which the reservoirs are supplied. The excavated plateau, extending from the foot of the mount to the precipitous bank of the river, is partly occupied by the wheel houses, forebays and other necessary structures, and the remaining spaces are very tastefully arranged, with flower gardens, gravel walks, fountains, statues and other ornamental devices, which, viewed in connection with the romantic country around, and the animating and busy scenes presented by the canals and rail-roads in the vicinity, form, altogether, a prospect of uncommon interest and beauty.

Previously to the erection of the works at Fairmount, the city had been supplied with water from the Schuylkill by means of two steam engines, one on Chestnut and Front streets, near the river, and the other at the intersection of Broad and Market streets. These were soon found to be wholly inadequate to the necessary supply, and were in a few years superseded by the works at Fairmount. By the first arrangement, the water was let into a basin, formed with suitable gates, at the foot of Chestnut street, and thence conveyed by an aqueduct, 460 feet in length to the water shaft at the lower engine house. Here it was raised by the engine and forcing pumps into a tunnel, 6 feet in diameter, extending along Chestnut and Broad streets, 3144 feet, to the other engine house at the Centre, now called Penn square. At this point, the water was again elevated, by the second engine, into a reservoir, 36 feet above the ground, and thence into an iron distributing tank, from which the wooden pipes, then in use, conducted the water through the various parts of the city. The total cost of this establishment from its commencement in 1799, to its abandonment in 1815, was $657,398 91, including $898 94 "*for whiskey;*" and the amount of water rents received during the same period, was $105,351 18, leaving a balance chargeable to the city treasury of $552,047 73.

In August, 1812, the construction of the steam works at Fairmount was commenced, and in September, 1815, was so far completed as to afford a partial supply of water to the citizens. In 1818, after expending $320,669 84 in the erection and support of these works, it became apparent that a more economical system, and one better calculated to secure the object in view, than the one then in use, must be adopted, and in compliance with a recommendation of the watering committee, councils immediately appropriated $350,000, and authorized the erection of the dam and other works, now in operation at Fairmount.

The dam, a mound of earth and stone, planked on its southern side, is 1600 feet in length, including the western pier, 150 wide at the base, 12 at top, and varying in height from 36 to 12 feet. The entire length of the overfall is 1204 feet, the eastern embankment 270, and the head arches through which the water flows into the mill race, 104 feet. At the western end of the dam is a short canal, with 2 guard, and 2 lift locks, constructed at the expense of the city, by agreement, for the use of the Schuylkill Navigation Company.

The strength of this dam has been subjected to many severe trials, but it has hitherto escaped serious injury. The great ice freshet of the 26th January, 1839, when the water rose 10 feet 2 inches above the top of the dam, and 12 feet 3 inches above high water in the river below, affected it more than any previous one. It completely inundated all the pump machinery, and by its force burst open the doors and considerably injured the partitions, floors, &c. of the mill houses, and carried away some of the planking and masonry of the dam.

The mill race forms a parallelogram, excavated from compact gneiss rock, to a mean depth of 38 feet, is 419 feet long, from north to south, 90 feet wide, and 6 feet deep below the overfall of the dam. It is bounded by a paved avenue, 253 feet long and 26 wide, and the mill houses on the west; on the east by the rocky and nearly vertical side of Fairmount, 70 or 80 feet in height, and on the north by the head arches, which are so constructed as to allow the passage into the race of a body of water 60 feet wide and 6 feet deep. By means of a waste gate, the water in the race may be drawn off and discharged into the river below the dam. The mill buildings are of stone, 238 feet long and 56 wide. The lower floor is divided into 12 apartments, 4 are intended for 8 double forcing pumps, of which six have been introduced. The other apartments are for the forebays leading to the water wheels. These wheels are all of the same length, but not of the same diameter, are formed of wood, having iron shafts weighing about five tons each. The pumps with a head equal in weight to 7900 lbs., force the water into the reservoirs at the top of the mount, 92 feet in height. The first of which was put in motion on the 1st July, 1822. It is 15 feet long and 15 feet in diameter, working under one foot head and seven feet fall. It forces one and a quarter millions of gallons of water to the receiving basin in twenty-four hours, with a stroke of the pump of four and a half feet, a diameter of 16 inches, and the wheel making eleven and a half revolutions in a minute. Five have since been put in operation, some of which make thirteen strokes in a minute, with small additional water fall, and force one and a half millions of gallons in twenty-four hours. Though the wheels are sunk below the ordinary line of high water, they are seldom affected except when the back water is about sixteen inches on the wheel.

The pumps are worked by a crank on the water wheel attached to a pitman connected with the piston at the end of the slides. They are fed under a natural head of water, from the forebays of the water wheel, and are calculated for a six feet stroke, but they are generally worked with not more than five feet. They are double forcing pumps, and are each connected with an iron main 16 inches in diameter, which is carried along the bottom of the race, to the foot of the mount, and thence up the bank into the reservoir, 92 feet above the dam.

The lowest estimate of the quantity of water afforded by the river in dry seasons, is 440,000,000 of gallons in 24 hours. The average quantity of water raised by each wheel and pump is about 530,000 gallons daily, but when the whole six wheels are put in motion, they can supply 6,000,000 of gallons in the 24 hours. The average daily consumption of water for the present year is about 4,000,000 of gallons, or 177 for each permit.

The reservoirs are lined with stone, and paved with bricks, laid upon a very tenacious clay bed, in strong lime cement, and made water tight. They are $12\frac{1}{4}$ feet in depth. The whole cost of the reservoirs was $133,824 42. From the central reservoir the water is conducted into the city by means of two iron pipes, one 20 and the other 22 inches in diameter. One passes down the north and the other down the south slope of the mount, each is nearly 10,000 feet in length ; additional mains have since been inserted in the same reservoir. In 1821, the work of laying down iron distributing pipes was commenced, and gradually displaced the old wooden pipes which had been used previously and exclusively. Of the 30 miles of wooden pipes laid from Fairmount through the city, in 1819, only 3 miles remain. Since the introduction of iron pipes there have been laid 62.62 miles of them up to January, 1840; add to which, 48.13 miles laid by the districts, and we have 109.75 miles. They extend about four miles in a south-east direction, and nearly the same distance towards the north-east. The larger iron pipes were originally imported from England; the whole cost of which, however, does not exceed $20,000; whilst those furnished by American manufacturers amount to $497,171 37.

The expense of supplying the city by steam power, with the same quantity of water now used, would be $206 a day; whilst the cost by water power, is $7 a day. This includes attendants' wages, fuel, light, &c. The estimated expenses for the year 1840, including general repairs and improvements, and extension of pipes, is $27,500; and the amount of the water rents for the same year, is $127,234 25; from which deduct the annual appropriation to the sinking fund, $17,000, and the estimated expenses for 1840, $21,209 67; making in all $38,209 67, and a balance remains, applicable to any other purpose, of $89,024 58. The whole sum expended at Fairmount since the employment of water power was determined on, up to December 31st, 1839, is $1,464,146 21; and the amount paid for salaries, labour, and incidental expenses, from 1812, is $379,428 19; making a total of $1,843,674 40. The amount of revenue derived from the city and districts for the use of the water, from the commencement of the works, is $1,493,024 53.

In addition to the innumerable pipes which convey the water into dwellings, &c., there are now distributed throughout the city and liberties 1007 "fire plugs," so called; to which, in case of fire, hoses, corresponding in calibre with the cavity of the plugs are attached, and thus convey the water to the engines, or, as is often the case, directly to the fire.

The average daily supply of water for the city and districts, during each quarter of the year 1839, was as follows:

PENNSYLVANIA. 109

	Gallons.
January, February, and March,	2,981,560
April, May, and June,	4,363,191
July, August, and September,	4,573,465
October, November, and December,	3,995,211

This shows an average daily supply for the year, of 3,978,357 gallons; and exceeds the consumption of the preceding year by 127,710 gallons.

In the city, the cost to each family supplied with water by private pipes, is $5 a year; the owner or occupant of the house paying all expenses of the introduction of the water into the premises. In the districts, each family pays $7 50 for the like supply. Hotels, manufactories, &c., pay an amount in proportion to the water supposed to be used, and generally at as high rates as families.

Their payments vary from $10 to $600 per annum. The County Prison pays $500; the City and Northern Liberties Gas-works, each $200; United States Mint, $85; stable keepers pay each $1 a year for each horse kept by them; hydrants for washing pavements, $2 each; small houses in the rear of other buildings, $2 50; and for openings, in private baths or lodging rooms, $3. Establishments similar to that at Fairmount, are now in successful operation in Richmond and Lynchburg, Virginia; Nashville, Tennessee; Cincinnati, Ohio; Wilmington, Delaware; Pittsburg, Lancaster, Allentown, and Bethlehem, Pennsylvania; the latter was established in 1752, and is probably the first work of the kind erected in this country.

UNION CANAL, extends from a point a short distance below Reading, to Middletown, on the Susquehanna, and passes through the counties of Berks, Lebanon and Dauphin. If the pool near Reading be regarded as a part of the Union Canal, that work commences about three miles below Reading, on the west bank of the Schuylkill, and running nearly due north, enters the valley of Tulpehocken creek; following that stream chiefly along its left bank, the canal gradually ascends to the summit, a distance of 41.29 miles. The summit level is 6.97 miles, and the western section, including ¾ of a mile of towing path, along the right bank of the Swatara, is 33.80 miles in length, making the entire length of the Union Canal, 82.08

miles. Course W. S. W. Summit at Lebanon, 498.50 feet above tide water; ascent 311; descent 208.50; total lockage 519.50 feet; 36 feet wide at top; 24 at bottom; 4 feet deep; 93 lift and 2 guard locks, each 75 by 8.50 feet; 43 waste weirs; 49 culverts; 135 bridges; 14 aqueducts; 1 tunnel, 729 feet in length. A navigable feeder from the Swatara, 6.75 miles long, and a pool formed by a dam at the head of the feeder, have also been constructed by the Union Canal Company. As the pool from which the summit is supplied is below the canal, the water is thrown into it by means of two forcing pumps, which are worked by water wheels; steam engines are provided for the same purpose, to be used in the event of accident to the wheels. This improvement affords a navigable communication from the main trunk of the Union Canal to Pine Grove, a distance of 23 miles, in a north-east direction. From Pine Grove the company have laid a rail-road through a gap in the Sharp Mountain, 4 miles in length, to the coal mines.

LEHIGH NAVIGATION. The Lehigh works, like those on the Schuylkill, consist of several canals and slack water pools. They extend from Easton to the Great Falls of the Lehigh, near Stoddartsville, in Northampton county.

Leaving the Lehigh immediately south of Easton, the line is conducted by locks, into the first canal, on the right bank of the river. About four miles above its point of outset, this section of the canal terminates at a dam one fourth mile below Smith's Island, and the pool thus formed is entered. This pool is about two miles in extent. At 'a distance of about six miles from Easton, commences the most extensive section of canal. It leaves the river one mile below Jack's Mill, passes along the north or left bank, through Bethlehem, and re-enters the Lehigh, at a dam not far from Allentown, in Lehigh county; thence by the river, one and a half miles; thence by canal to a dam, three miles; thence to river, three-quarters of a mile; thence by canal to a dam, four and a half miles; thence by the river, one and a half; thence by canal one mile to a dam; thence by the river, two and a half miles; thence by the canal three and a half miles to the Lehigh Water Gap. Here the Canal passes the Great Blue Mountain, and enters the coal region. From the termination of the last mentioned canal, the

PENNSYLVANIA. 111

river is used for one mile to a dam at the mouth of the Aquanshicola creek; then occurs a canal four and a half miles long, to a dam; thence by the stream one mile; and then is entered the canal, six miles in length, which extends to Mauch Chunk. From Mauch Chunk to Whitehaven, 24.75 miles, the improvements consist of canal and slack water navigation, similar to that below Mauch Chunk. From Whitehaven to Wright's creek, about one and a quarter miles, it is slack water; and thence to the Great Falls, at Stoddartsville, it is for a descending navigation by artificial freshets. The works from Easton to Mauch Chunk are 46.23 miles in length; and from Mauch Chunk to their northern terminus, 38.25 miles. Total length, 84.48 miles; of which 30.53 miles consist of pools; 39.26 of canals; 2.48 of locks; and the remainder of sluices.

The canals above Mauch Chunk are 60 feet wide at top water line, 40 feet at bottom, and 5 feet deep.

The locks, 29 in number, are each 20 feet wide, 100 between the quoins; 86 feet clear of the swing gates; 10 to 30 feet lift; and are capable of passing boats of more than 100 tons. One of the locks has a lift of 30 feet, which is filled or emptied in two and a half minutes.

High water guard, 5 to 6 feet. Working guard, 3 to 4 feet. Twenty dams, from 187 to 375 feet long, and from 14 to 38 feet high. Total fall, 935.83 feet.

The canals at and below Mauch Chunk, are 60 to 65 feet wide at top water line, 45 feet at bottom, and 5 feet deep. Five guard, 3 guard and lift, and 44 lift locks, 22 feet wide, 100 between the quoins, 85 feet clear of the swing-gates; 6 to 9 feet lift; pass boats carrying more than 100 tons; 8 dams from 300 to 564 feet long, and 8 to $19\frac{1}{2}$ feet high. Total fall, 353.2 feet.

The export of coal by the Lehigh Company, during the year 1839, was 142,507 tons; and by other companies, 79,343 tons; total conveyed on the Lehigh Canal, 221,850 tons. In 1837 the Lehigh Company sent down the canal 200,000 tons.

The tolls received on 273.190 tons of coal and other articles amounted in 1839 to $141,300 11.

This company's coal lands, amounting to six thousand acres, comprise the whole of the east end of the first or southern anthracite coal field, beginning on the top of the mountain, about half a mile from the Lehigh river, and near Mauch Chunk, and extending without interruption to Tamaqua, on the Little Schuylkill, a distance of 13 to 14 miles. On these lands are found, beginning on the north side of the Coal Basin, nine veins from 5 to 28 feet in thickness, making together 111 feet. On the south side, which has not been so fully examined, are found veins of 50, 20, 15, and 9 feet. This coal is now penetrated, from the Room Run Valley, which cuts into the mountain on the northern side of the Coal Basin, and near to its base, and thus exposes the veins above-mentioned. At the Old Mine, five miles west of Room Run, the vein of 50 or 60 feet, which is the only vein worked at this place, lies as a saddle on the top of a hill nearly as high as the main mountain; here the coal is removed by quarrying in open day. About 30 acres have been worked out from this single vein, which have produced upwards of 1,100,000 tons.

Connected with the Lehigh Navigation, are several railroads leading from the various coal mines, situated in what are termed the first and second coal fields, whence large quantities of anthracite coal are sent to Philadelphia by the Lehigh and Delaware Canal, and to New York by the Morris and Delaware and Raritan Canal. Among these are the Beaver Meadow; Hazelton; Nesquehoning; Wilkesbarre; Mauch Chunk; Buck Mountain; Sugarloaf, and other small rail-roads.

LACKAWAXEN CANAL, see Hudson and Delaware Canal, New York.

CONESTOGA CANAL, consists of dams and locks. It commences at Reigart's landing in the city of Lancaster, and terminates at Safe Harbour on the Susquehanna. Length, 18 miles; course, south-west; 9 locks, each 100 by 22 feet; 9 dams; descent 62.

CODORUS NAVIGATION, is similar to the preceding, the improvement having been effected by means of canals and pools. It extends from York to the Susquehanna river and consists of 8 miles of slack water pools, and 3 of canals; length 11 miles; course, north-east; 9 locks.

BALD EAGLE AND SPRING CREEK NAVIGATION, extends from the state dam, on the Bald Eagle Creek, at the head of the side cut, to the town of Bellefonte in Centre county. Length 25 miles; cost so far as finished, (19 miles) $230,000. This improvement, though under the control of a joint stock company, is in fact a state work, constructed on the faith of the commonwealth which is pledged for the payment of an interest of 5 per cent. per annum for 25 years on $200,000, and has since become a stockholder to the amount of $25,000.

WEST PHILADELPHIA CANAL, is a small canal around the western abutment of the bridge over the Schuylkill, near Philadelphia. It is designed for the use of such vessels as cannot pass underneath the bridge, and enables those engaged in the coal trade to approach the first lock of the Schuylkill Navigation. Length one-twelfth of a mile.

SUSQUEHANNA CANAL, commences at Wrightsville, opposite Columbia on the Susquehanna river, and descends the right or west bank of that stream to Havre de Grace in Maryland. Length 45 miles; 50 feet wide at top; 5 feet deep; 29 lift and 2 guard locks, double chamber, and admit the passage of two boats each 85 feet long at the same time, or 1 raft 170 feet long and 16 wide; total lockage 233 feet. This work, sometimes called the "Tide water Canal," opens a communication between the Central Division of the Pennsylvania Canal and Chesapeake Bay. In structure, it is similar to the Pennsylvania Canal, and is designed as a continuation of that work, to tide water, though owned by a private company.

CHESAPEAKE AND OHIO CANAL. See Maryland.
SANDY AND BEAVER CANAL. See Ohio.
MAHONING CANAL. See Ohio.

RAIL-ROADS.

COLUMBIA AND PHILADELPHIA RAIL-ROAD, the first link in the great western chain, commences at the intersection of Vine and Broad streets, Philadelphia, pursues a western course, and terminates at Columbia on the Susquehanna. Length 81.60 miles. This rail-road opens a direct communication between the valleys of the Delaware and Susquehanna, and intersects those of the Schuylkill, Brandywine and Conestoga, passing through the counties of Philadelphia, Chester and Delaware; and the towns of Downingtown, Lancaster, &c. The West Chester branch leaves the main line at a point 22 miles from Philadelphia, and that to Harrisburg, in the city of Lancaster.

This road forms a part of the great thoroughfare to Pittsburg and the western states, and is the most important outlet of the city of Philadelphia, towards the valley of the Mississippi. At its point of termination at Columbia, commences the Central Division of the Pennsylvania Canal which, with the Allegany Portage Rail-road, and the Western Division of the Pennsylvania Canal, completes the "Rail-road and canal route to Pittsburg." An extension of this road from Columbia to York, in York county, is nearly completed; and a farther extension towards Gettysburg was advancing, when an order from the legislature, during the session of 1838–39, arrested its further progress. The Columbia Rail-road is the property of the commonwealth of Pennsylvania; the legislature of which authorised its construction on the 24th of March, 1828, and its location soon followed. On the 20th of September, 1832, twenty miles of single track were ready for use; in April, 1834, a single track along the entire route from Philadelphia to Columbia, was opened for travelling; and in October of the same year, the second track was completed, and the road opened for public use. The depots, work-shops and other necessary structures, were subsequently completed.

At a distance of about two miles from its point of outset, the road crosses the Schuylkill by a viaduct 984 feet in length, and immediately ascends an inclined plane of 2805 feet in length and 187 feet in height; and thence pursues its course along the dividing ridge between the Delaware and Schuylkill to a point near the intersection of the West Chester Rail-road, where it attains an elevation of 543 feet above high tide. Hence the road descends the South Valley hill into the great Chester valley, to Downingtown; from the summit of the South Valley hill to the Big Brandywine bridge, which is 250 feet above tide, the descent is at the rate of 29 feet per mile. After crossing the Little Brandywine, the road ascends the North Valley hill until it attains the summit at Mine Ridge Gap. Here the soil, being such as to forbid the excavation of 37 feet as originally intended, it was determined to increase the grade so as to reduce the depth of excavation to 23 feet; the grade, therefore, from the summit on both sides now stands at 45 feet per mile, for three-fourths of a mile, and thence a farther distance of one-fourth of a mile at 40 feet, when it resumes the original inclination of 30 feet per mile. From the Gap summit, which is 533 feet above high tide at Philadelphia, it proceeds through Lancaster, and enters Columbia at the outlet lock of the Pennsylvania Canal. The plane by which the town of Columbia was formerly entered, is 1800 feet in length and 90 in height.

After many vexatious delays, occasioned by individuals, whose personal interests were likely to be affected by the location of the road, especially its eastern section, Major Wilson, the efficient engineer in chief, proceeded to the execution of his important task. Having determined upon its route, the principles of its construction next engaged his attention. The maximum grade of the line was fixed at 30 feet per mile and its minimum radius of curvature 631 feet. These principles were rigidly adhered to, with the trifling exception at Mine Ridge Gap, above-mentioned. As the inclined planes augment the expense and time of transit on this road, efforts have been made to avoid them. A new route of six miles has been completed, by which that at Columbia is dispensed with; the distance is nearly the same as the abandoned section, but its grade is 35 feet per mile. Several routes have been surveyed for the purpose of avoiding the inclined plane near Philadelphia; but as yet no alteration has been made. Two roads for this purpose have been commenced by joint stock companies; the West Philadelphia Rail-road, about 8 miles in length, with a maximum grade of 57, and an average grade of 43.30 feet per mile, and the Valley and Norristown Rail-roads; by the latter, the distance to Columbia will be increased 2.12 miles.

There are on the Columbia Rail-road nearly 57 miles of straight line; 12 miles with a mean radius of 2230 feet, and the remainder with that of 822 feet. The width of the road is 25 feet in the excavations, the top width of embankments generally exceeds 25 feet. The deepest cuttings are between 30 and 40 feet, and the highest embankment is 80 feet. A building at the head of the Schuylkill inclined plane, contains a stationary steam engine of 60 horse power. The rope used for elevating the carriages, is an endless one, 9 inches in circumference when new, and cost about $2,800. The first rope used was 6.75 inches in circumference, and cost $2,100, weighed

5.25 tons, and lasted about one year. On this plane cars pass up and down at the same time.

The culverts, 75 in number, are built of stone, and the masonry is either hammer or rubble work, with spans, varying from 4 to 25 feet, and contain 31.161 perches of masonry.

The number of viaducts is 20; they are constructed with stone abutments and piers, surmounted by wooden structures. There are 33 bridges across the rail-way for public and private roads.

The superstructure of the Schuylkill viaduct is of wood, with distinct trusses, formed of arch piers, king-posts and braces. The whole width from out to out is 49.67 feet, which admits of three separate passages, two of 18.50 feet each, in the clear, and one of 4 feet; the latter is used for foot passengers; one of the former for two rail-way tracks, and the other for common carriages. The spans are seven in number and six piers. The whole length of wooden platform is 1,045 feet, and the height of bridge floor above usual water line, is 38 feet. The total cost was $133,947.

Valley creek viaduct has four spans, each 130 feet in clear between the piers. Piers vary from 56 to 59 feet in height. Cost, including stone work, $22,254. The wood work was recently destroyed by fire, and replaced by a lattice bridge, (depressed so as to admit of the rail-way being carried over the top.) Cost, $17,218.

East Brandywine viaduct, four spans, two of 88.66 feet each, and two of 121.58 feet in the clear. Clear width 18.50 feet; length of platform, 477 feet, and height of floor above water, 30 feet. Cost, $17,523.

The West Brandywine viaduct, has a wooden superstructure, resting upon stone piers and abutments. Length of platform, 835 feet, divided into six spans; its greatest height above the water is 72 feet. The whole of stone and wood work. Cost, $57,916. In this, like the one over Valley creek, the line is carried over the top.

Big Conestoga viaduct, is 1412 feet in length, and is elevated 60 feet above the water; stone piers and lattice superstructure on Town's plan. Cost, 31,503. The longest span of the bridge is 120 feet.

Little Conestoga viaduct, stone piers and abutments; flooring

804 feet in length; elevation 47 feet above the water. Cost, $15,359.

Mill Creek viaduct, length of platform 550 feet; elevation above the water 40 feet. Cost $9,273.

Pequea viaduct, single span 130 feet; cost $8,735. This, like most of the others, is on Burr's plan.

Railway superstructure.—The entire length of *single track* is 163.20 miles, 6 miles of which have granite sills, plated with flat iron bars; 16 with wooden string pieces, similarly plated; 2 miles with stone blocks and edge rails, having stone sills, extending across the track at intervals of 15 feet; and 137.20 miles with stone block and edge rail, having wooden sills across the track, except on some of the embankments, where the edge rail is secured to cross sills of wood, supported by mud sills.

Granite track.—The trenches are dug in the direction of the road, two feet wide and 22 inches deep, measuring from the level of the top sill. Broken stone is then placed compactly, in layers of 3 inches each. Upon this are laid granite sills varying in length from 3 to 12 feet, and one foot in depth and width. Holes are drilled into the stone, 3.50 inches in depth and 5-8 of an inch in diameter. Into these holes, plugs of locust wood are driven, to receive the spikes which secure the iron bars, which are 15 feet in length, 2.25 inches wide and 5-8 of an inch in thickness. The inner edge of the sill is chamfered off for a width of two inches, and the outside is backed up with broken stone. Horse power being used on the road when this track was laid, a horse-path was formed of broken stone or gravel 6 inches in depth. The average cost of one mile of this track, including the trimming and dressing off half the width of the road-way, was $10,179 20.

Wooden track.—The trenches are dug across the road, four feet apart, eight feet in length, one foot in width and 16 inches in depth, (making 24 inches to top of wooden rail.) Into these, broken stone is rammed in layers, upon which are laid sills of chesnut or white oak, 7.50 feet long and 7 inches square. The sills are notched to receive a yellow pine string piece, 6 inches square, which is secured in its place by wooden wedges. Flat iron bars are then spiked on, similar to those used on the granite track; the horse-path is also similar. This track cost $5,604 48 per mile.

The two kinds of structure just described, have been in use about seven years, during which the wooden sills and string pieces have become much decayed; some of the bars also are broken and displaced, and in consequence they are working loose. This part of the road is to be renewed with edge rails,—a portion of which are already laid.

Edge rails on stone blocks and sills.—The trenches are dug in the direction of the road, 28 inches wide and 24 deep from top of block; at 15 feet these are connected by a cross trench, 16 inches wide. Broken stone to the depth of 12 inches, is well rammed in layers; the blocks and sills are then settled in their places by heavy rammers, and backed up to their tops with broken stone. The blocks are of granite or other hard stone, 20 inches long, 16 wide, and 12 deep; the sills are of the same material, 6.50 feet long and one foot square, placed across the track at every 15 feet; the blocks are so arranged as to give support to the rails at every three feet. Cast iron chairs, weighing 15 lbs., are secured to the blocks and sills, by bolts driven into cedar plugs previously inserted into the stone; there are two bolts to a chair, weighing 10 ounces each; between the stone and chair, a piece of tarred canvass is inlaid. The rails are of rolled iron, 15 feet long, 3.50 inches deep, parallel at top and bottom, and weigh 41.25 lbs. per lineal yard. The rail is secured in the chair by two wrought iron wedges, one on each side, weighing 10 oz. The horse-path for this track is formed of broken stone and gravel, 9 inches deep. Average cost of one mile, $12,568 85.

Several miles of track were laid in a similar manner to the above, omitting the stone sill, and substituting in its place two blocks, at a cost of $10,927 88 per mile. This kind of track was found so liable to spread, particularly in the spring of the year, that wooden sills have since been put in at intervals, connecting the two rails of the track.

Edge rails on stone blocks and locust sills.—This kind of track is similar to the edge rail track already described, with the following exceptions; instead of stone, locust sills are used, placed 15 feet apart on the straight lines, and 9 feet apart on the curves; to suit which, some bars were rolled in lengths of 18 feet; the stone horse-path is dispensed with, the tops of the blocks and sills being level with the graded surface of the road.

The average cost of one mile on this plan is $13,249 92 ; the excess over the cost of the track in which stone sills were used, is owing to a rise in the cost of iron, from $41 to $50 per ton, (delivered at the eastern end of the road.)

On newly formed embankments the following plan was adopted : longitudinal trenches were dug, 22 inches wide, and 22 inches deep ; broken stone to the depth of 6 inches, being rammed in, string pieces of white oak or chesnut were laid, 12 inches deep by 10 inches wide ; these being notched to the depth of two inches, cross sills of the same material, 6 by 8 inches, were secured to them at every 3 feet by pins or wedges. On these sills the iron chairs, rails, &c. were placed. The trenches were connected at intervals, by cross trenches, running out to the edge of the embankment, for the purpose of carrying off the water. This description of track cost $12,905 35 per mile. This road having been originally constructed for horse power, a system of turn-outs and side-tracks was adopted. Turn-outs were placed at intervals from one track to the other, and side-tracks were laid, adjacent to each of the main lines, at the distance of one mile and a half apart, for the whole length of the road ; these side-tracks measured as follows :—160 feet in length parallel to the main track, and 70 feet at each end, curved to the intersection with the outside rail of main track. They afforded a space of about 200 feet in length for cars, and as the cars always entered in the same direction after both tracks were completed, only one moveable switch was used. Upon the introduction of steam power, the old castings having been found objectionable, were displaced, and others better adapted to this object, laid down ; most of the side-tracks were also removed.

The following table exhibits the cost of the Columbia Railway, as nearly as can be ascertained. It must, however, be borne in mind, that since the road was opened for public use, various sums have been appropriated to it, in addition to previous appropriations ; some portions of which belong, properly, to the item of construction, while others have been applied to objects not connected with its construction.

Total cost of the Columbia and Philadelphia Railway.

Grading,	$649,158 69
Culverts,	74,113 94
Viaducts or rail-way bridges,	327,695 80
Road and farm bridges,	42,055 00
Fencing,	65,410 86
Rail-way superstructure,	2,181,156 25
Building and machinery,	111,787 12
Engineering and superintendence,	133,934 31
Damages,	54,833 29
Repairs,	42,451 76
Incidental,	11,980 18
Alteration to accommodate the city of Lancaster,	60,000 00
	$3,754,577 20

Since the road was opened in 1834, the following items of expenditure are to be added:—

Locomotive engines,	$327,203 41
Additional buildings, turn-outs, &c.	37,511 16
Retained per centage on former contracts,	5,134 08
Engineering,	4,741 25
New ropes at inclined planes,	11,584 34
Embankment at Maul's bridge,	1,796 34
Renewal of wooden track,	18,907 48
Rebuilding Valley Creek bridge destroyed by fire,	17,218 13
New road to avoid Columbia inclined plane,	118,123 53
Grand total	$4,296,796 92

The total expenses of working the road for one year, commencing October 31st, 1837, to October 31st, 1838, were:—

Ordinary road expenses,	$44,033 23

PENNSYLVANIA. 121

Motive power	"	.	.	133,820 99
				$177,854 13

Receipts during the same period.

Road tolls,	.	.	.	$233,588 75
Motive power tolls,	.	.	.	164,052 74
				$397,641 49
Deduct expenses,				177,854 13
Profit to the state,				$219,787 36

During the year just mentioned there were 103,336 passengers and 87,180 tons of merchandize conveyed upon the road. All the cars used on the road belong to individuals or companies, but the motive power is furnished by the state. Horse power is used on the West Chester Rail-way and a few others.

The officers and attendants of the road consist of one "superintendent of motive power," who has charge of every thing in that department. One supervisor, who is charged with the repairs, &c. These officers are wholly independent of each other; they appoint all persons employed under them, respectively, and report annually to the Board of Canal Commissioners, by whom the collectors of tolls, five in number, are appointed.

The rates of toll, vary from 6 mills to 4 cents per ton (of 2000 pounds) per mile; there are twelve different rates, the average of which would be 2 cents per ton per mile. The lowest rates are for coal, stone, iron, vegetables, lime, manure, and timber, and the highest are for dry goods, drugs, medicines, steel and furs. On the United States mail, the toll is one mill per mile for every 10 pounds; on every passenger, one cent per mile. In addition to these rates, a toll is levied of one cent per mile on each burthen car, two cents on each baggage car, and on every passenger car, one cent per mile for each pair of wheels. The motive power toll is, for each car having four wheels, one cent per mile, for each additional pair of wheels five mills, for each passenger, one cent per mile, and for all other kinds of loading, 12 mills per ton (2000 pounds.) The owners of cars now charge $3 25 for each passenger, and $7 50 for every ton of merchandise conveyed the whole length of the road, they paying all tolls; which is at the rate of 4 cents per mile for a passenger, and $9\frac{14}{100}$ cents per mile for a ton of goods.

The heavy locomotives now used for the transportation of freight, are capable of drawing thirty-five cars, each with a load of three tons, or one hundred and five tons, exclusive of the cars, engine and tender; if these be added, the whole will be 190 tons. The number of locomotives on the road at the date of the last report, was thirty-six, of which twenty-seven were in good order. The daily duty of the engines is to run about seventy-seven miles. During the year 1839, 51,156 cars passed ever the Schuylkill plane, and 52,664 over that at Columbia.

Subjoined is an article relative to the cost of motive power, which is from the American Rail-road Journal, and which, in connection with the preceding remarks will afford a satisfactory view of the whole subject.

"In 1838 the cost of motive power, for repairs, oil, fuel, attendance, &c., was per mile run on the

Boston and Lowell Rail-road, . . . 94 cents.
Boston and Worcester Rail-road, . . 79 "
Baltimore and Ohio Rail-road, . . . 1 60 "
Richmond and Fredericksburg Rail-road, . 80 "
Philadelphia and Columbia Rail-road, . . 55 "

The length, and the manner in which each of these roads is built, and the kind of engines used on them, are all before the world, and we presume our readers are familiar with their history; it is therefore unnecessary to make any remarks with regard to them. It is also well known that the Philadelphia and Columbia Rail-way is owned by the state of Pennsylvania, and the motive power is supplied by the state, while the cars are owned by individuals or companies. In making a statement of what profit the road would have given to the state, if it had owned the cars, we will assume an indebtedness for them in addition to the cost of road and motive power, when we shall find that it paid a profit upon the whole outlay, of *nearly* $12\frac{1}{2}$ per cent.

PENNSYLVANIA.

Original cost of the road,	$3,333,236
Fifty locomotive engines cost	336,000
Various appurtenances,	330,764
Cost of passenger depots, supposed,	200,000
Pay of agents and officers,	55,625
Three hundred and sixty-three cars at $275 each,	99,825
Twenty passenger cars at $2,000	40,000
Wear and tear,	27,964
Contingencies,	20,000
	4,443,414

In the year 1838 there was carried over the
road 87,180 tons 82 miles at $7 50 per ton, 653,850 00
75,612 passengers, $3 25,   .   245,739 00

$889,589 00

The expenses were for carrying
  87,180 tons at $2 50,   . $217,950
The expenses were for carrying
  75,612 passengers at $1 60, 120,979 80   338,929 80

Net receipts, $550,659 20

Which is 12.39 per cent on the preceding statement of cost. We consider it as very remarkable that the state can manage a road with more profit than a company; yet so it is; and as some may doubt the correctness of the assertion, we give the different expenses in detail, which are as follows:

A statement of the cost of working the Philadelphia and Columbia Rail-road, from October 31st, 1837, to October 31st, 1838.

Cost per trip, the distance of 82 miles,	$44	03 c.	5 m.
The fuel costs per trip, of 82 miles,	14	04	1
Cost per ton, the distance of 82 miles,	1	55	3
Cost per ton per mile, 7,562,040 tons,		1	8
Fuel cost per ton, 82 miles,	.	50	79-1000
Cost of repairs per ton, 82 miles	.	27	4
Cost of repairs per ton per mile,	.	3	3

Cost per mile travelled for repairs of engines,	9 c.	7 m.
Cost per mile travelled 260,400, including all repairs, attendance, &c.,	54	99-100
Cost of maintenance of planes per ton, 82 miles,	18	3
Engineer's and firemen's pay per ton, 82 miles,	18	8
Cost of maintenance of planes per mile per ton,	2	2
Engineer's and firemen's pay per ton per mile,	2	3
Cost for fuel per mile travelled,	13	86-100
No. of tons per trip way and through, 28 1-5 useful load,		
No. of cars per trip 14 2-7.		
Cost of oil per ton per trip, 82 miles,	7	1
Cost of oil per ton per mile,		8
Cost of oil per mile travelled,	2	5 2-10
No. of tons through and way trains, useful load 42 1-7		

Total number of tons hauled, allowing 15 passengers to a ton, and 87,180 tons of merchandise, was 92,204 tons 82 miles, as copied from the book of performances kept in that year.

A statement of the work done on the Philadelphia and Columbia Rail-way by 13 engines, manufactured by M. W. Baldwin, and the cost; said engines being taken in order as they came on the road, being the 13 last furnished by him to the state, from the time they commenced running till 31st October, 1838. [See Table next page.]

PENNSYLVANIA.

1837. When commenced.	Class.	No. of miles travel-led.	No. of cars haul'd.	No. of ts. dis. 77 ms. 3 ts. pr car.	No. of ts. dis. 1 m. over as- cent of 45 ft per mile.	No. of trips.	No. of ts. per trip thro'	Cost of repairs to engines.	Cost pr. m. pr. haul'd car.	Cost pr. ton haul'd ton prm.	Cost pr. ton dis. 77 miles.
Westchester, Feb. 19.*	3d	30.636	1.973	5.919	455.763	268	22.08	1.715.97	5c 6m	.76c	28.97
Virginia, Feb. 19.†	"	36.421	3.729	11.187	861.399	473	23.65	1.658.48	4.55	1.92	14.82
Paoli, Feb. 19.‡	"	36.036	3.426	10.278	792.099	468	21.98	1.148.45	3.16	1.44	11.14
Connestoga, F'b. 22.‡	1st.	5.929	1.549	4.647	357.819	77	60.35	131.62	2.21	.36	2.83
Ed. F. Gay, March 24.	"	25.872	7.265	21.795	1.678.215	336	64.86	1.475.78	5.63	.87	6.68
Parksburg, April 2.	"	24.178	6.361	19.083	1.469.391	314	60.77	1.591.29	6.58	1.08	8.33
Octarara, April 7.	"	13.552	3.628	10.884	838.068	176	61.84	771.90	5.69	.91	7.09
Pequa, April 24.	"	14.168	3.664	10.992	846.384	184	59.73	1.221.69	8.61	1.44	11.11
Downingtown, Ap. 16.	"	26.257	7.074	21.222	1.634.094	341	62.23	1.475.23	5.64	.9	6.95
Indiana, May 1.	"	26.026	6.975	20.995	1.611.225	388	61.90	562.80	2.16	.34	2.68
Mississippi, May 9.	"	15.323	3.915	11.745	904.365	199	59.02	1.384.01	9.04	1.41	11.07
Montgomery, May 15.	"	21.406	5.261	15.783	1.215.291	278	54.99	830.64	3.88	.68	5.32
Wisconsin, May 28.§	"	8.624	2.160	6.400	480.960	112	51.85	82.22	.95	.17	1.26
		274.428	56.980	170.940	13.162.380	3.564		14.031.59	5.18 avera.	1c. avera.	9. average.

* This was run 10,000 miles, below the Schuylkill plane, of which the number of cars was not kept. † Ran the passenger train.
‡ This was on the Portage road six months. § This was used on a ferry boat, at Clark's ferry, all the season.

N. B. All those engines whose repairs exceed $1000, met (during the period of seventeen months, at different times,) with accidents, such as running off the track, and breaking their axles, springs or frames, so that the mere wear alone, or repairs occasioned by running, would have been less. The West Chester is not allowed any cars or expenses for 10,000 miles which she run from Broad-street to the Schuylkill plane —all her repairs being charged to the number of cars she hauled over the road, which, if allowed, would diminish her expenses considerably.

The Paoli and Virginia, run with passenger trains, took less cars, but run more trips—the first running 473 out of 530 working days; the second 468 out of the same number of days. One losing 57 days, the other 62. The other engines did not fill up the time so, because freight was not to be had at all times."

ALLEGANY PORTAGE RAIL-ROAD. This work commences at the termination of the Central Division Pennsylvania Canal, at Hollidaysburg, pursues a W. N. W. course to Blair's Gap, and thnce turning to the S. W. enters and passes along the valley of the Connemaugh to Johnstown, in Cambria county, having traversed in its course portions of Huntingdon, Bedford and Cambria counties. This road is connected with the central and western divisions of the Pennsylvania Canal, by two extensive basins, with which it communicates with slips and branch rail-ways. Length 36.69 miles; rise from Hollidaysburg to the summit 1398.71 feet, in a distance of 10.10 miles; and fall from the summit to Johnstown 1171.58 feet, in a distance of 26.59 miles; total rise and fall 2570.29 feet; of which 2007.02 are overcome by planes, varying in inclination from 4° 9' to 5° 51', or from 7.25 feet to 10.25 feet elevation, to 100 feet base. The planes are all straight in plan and profile; and 563.27 feet by grading. With the exception of the ends, the grades never exceed 21.12 feet, and are generally between 10 and 15 feet, per mile. Aggregate length of the bases of the inclined planes, 4.37 miles, and that of the graded portion of the road, 32.32 miles. The embankments are 25 feet wide on the top. There are four extensive viaducts; one over the Connemaugh at the Horse Shoe bend, which is a magnificent structure, with a single arch of 80 feet span, and the top of the masonry is 70 feet above the surface of the water. The cost of this work was $54,562 24. One at the Ebensburg branch; one at the Mountain branch; and one across the Beaver Dam branch of the Juniata. Of culverts there are 68; 85 drains; and several bridges; 11 levels; 10 inclined planes, 5 on each side of the mountain; 1 tunnel, about four miles from Johnstown; it is 901 feet long, and 20 feet wide by 19 high within the arch; cost of tunnel $37,498 85. Width of the road 25 feet, exclusive of side drains.

The edge rails used on the Allegany Portage, are " parallel " rails of rolled iron, weighing about forty pounds per lineal yard. They are supported by cast iron chairs, which weigh on an average about thirteen pounds each. The rail is secured in every chair by one iron wedge. The stone blocks which support the chairs, contain three and a half cubic feet each, and they are imbedded in broken stone, at a distance of three feet from centre to centre. On a part of the rail-way, the chairs are laid upon a timber foundation; and on the inclined planes, and along the canal basins, at the two terminations of the road, flat rails upon timber are used. At the head of each inclined plane, there are two stationary steam-engines of about thirty-five horse power each, which give motion to the endless rope, to which the cars are attached. Only one engine is used at a time, but two are provided to prevent delay from accidents. Four cars, each loaded with 7000 lb. can be drawn up, and four may be let down at the same time; and from six to ten such trips can be made in an hour. A safety car attends the cars, both ascending and descending, and stops them in case of accident to the rope, which adds greatly to the security. The grubbing and clearing of the Portage Rail-road cost $30,524. This work was equal to cutting a road through a dense forest, 120 feet wide and about 30 miles long. The grading of the rail-road, including the grubbing and clearing, and all work done under the contracts for grading cost $472,162 59¼. This work includes,

337,220 cubic yards of common excavation.
212,034   "   slate or detached rock.
566,932   "   hard-pan or indurated clay.
210,724   "   solid rock.

14,857 cubic yards of solid rock in tunnel, at $1 47.
967,060 " embankments carried over 100 feet.
67,327 perches slope-wall, of 25 c. feet.
13,342 " vert. " and wall in drains.

The viaducts and culverts, and the skew bridge for carrying the turnpike over inclined plane No. 6, contain 28,368 perches masonry, and their total cost was $116,402 64¼. For the first track and the necessary turn-outs, including a double track upon the inclined planes, there were delivered 50,911 stone blocks, each containing three and a half cubic feet, cost $27,072 15; and 508,901 feet lineal of 6 by 8 inch timber; 239,397 feet of 10 by 10; and 2,842 feet of 12 by 12 inch timber, of white oak and pine, which cost $47,184 50. The work done under the contracts for "laying" rail-way on the first track, including furnishing broken stone, amounted to $135,776 26. The total cost of British iron at Philadelphia imported for the first track, was $118,888 36. The *aggregate cost* of all the work done and materials furnished under contracts for the *first track* of rail-way, was $430,716 59½. For the second track there were imported 16,976 bars of edge rails, each eighteen feet long, which weighed 1803 tons, 14 cwt. gross, and cost at Philadelphia $87,494 80, or $48 51 per ton. The *aggregate cost* of all work done, and materials furnished under contracts for the *second* track of rail-way was $362,987 05½. Aggregate cost of work done and materials furnished under contracts for building ten stationary engines and machinery at the inclined planes, houses, sheds, dwelling-houses for enginemen, wells, water-pipes and ropes, first set, was $151,923 30¼.

General statement of the cost of Portage Rail-road.

Cost of Grading, . . - .	$472,162	59¼
Masonry, , . . .	116,402	64¼
First track of Rail-way, .	430,716	59½
Second " " .	362,987	50½
Buildings, Machinery, &c. at planes, *first set,* . . .	151,923	30¼

Ten Stationary Engines, *second set*,	37,779	25
Buildings, &c. for second set of engines,	21,048	59
Depots, Machine Shops, Water Stations, Weighing Machines, and various works,	41,336	66½
	$1,634,357	69¼

The above sum is the cost of constructing the Portage Rail-road at the contract prices; but it does not include office expenses, or engineering, or the extra allowances made to contractors, in a few instances, by the legislature after the work was completed, and beyond the contract prices.

Four locomotive engines have been used upon the " long level," but the expenses of them belong to another account.

In its course from Hollidaysburg to Johnstown the road attains an elevation of 2,491 feet above the Atlantic ocean. At Johnstown the Portage Rail-road joins the Western Division of the Pennsylvania Canal.

The execution of this important work was authorized by an act of the legislature of Pennsylvania, passed on the 21st of March, 1831. It was commenced on the 12th of April, 1831, and completed March 18th, 1834.

PHILADELPHIA RAIL-ROADS. The various important rail-roads which concentrate at Philadelphia, are extended into the city and surrounding districts by several minor works; among which are the following:

CITY RAIL-ROAD, commences at the termination of the Columbia Rail-road, at the intersection of Vine and Broad streets; extends thence down the latter, and terminates at the crossing of Cedar or South street, where it unites with the Southwark Rail-road. Length 1 mile; double track.

MARKET-STREET BRANCH OF THE PRECEDING, leaves the main line at Broad street; proceeds eastward to Third street; thence south, to, and along, Dock street, to an extensive range of buildings, erected by the city authorities, for the accommodation of the tobacco trade. Length 1.25 miles; double track to Eighth street; thence to Dock street, single track; and along Dock street double tracks are laid. These roads are the property of the city.

NORTHERN LIBERTIES AND PENN TOWNSHIP RAIL-ROAD, diverges from the Columbia Rail-road near its eastern terminus, and proceeds eastward, along James and Willow streets, to the Delaware. In its course it is intersected by the Germantown and Norristown, and Philadelphia and Trenton Rail-roads. Length, about 1.25 miles; constructed by a joint stock company.

SOUTHWARK RAIL-ROAD, extends from the termination of the City Rail-road at Broad and South streets, along the former, to Prime street, where it is intersected by the Philadelphia and Wilmington Rail-road. After uniting with that road, it curves towards the east, and proceeds down Prime street to the Delaware, near the Navy Yard. Length, about two miles.

A branch leaves this road and extends, along Swanson street, to South street. Length, half a mile.

WEST PHILADELPHIA RAIL-ROAD, Pa., commences on the Schuylkill river about 400 feet below the Market-street bridge, pursues a north-west course, through Hamilton and Mantua villages, and joins the Columbia and Philadelphia Rail-road about 8 miles from the Schuylkill. This road is intended to avoid the inclined plane on the latter at Peters's Island. Its maximum grade is 57 feet per mile, and its average grade 43.30 feet per mile. The grading of this road is principally done, but the work is now suspended.

VALLEY RAIL-ROAD, extends from Norristown to a point on the Columbia and Philadelphia Rail-road, about 31 miles west of Philadelphia. It is 20.25 miles in length, with a maximum inclination of 35.70 feet per mile. From Norristown the road has ascending grades nearly its entire length.

WEST CHESTER RAIL-ROAD, connects the Columbia Rail-road with the village of West Chester; it commences on the South Valley Hill, 22 miles from Philadelphia, and pursues the general course of the ridge about 10 miles to West Chester, in Chester county. Single track, though graded for two tracks. Width between the tracks is 4 feet, $8\frac{1}{4}$ inches. Maximum radius of curvature, 1260 feet; minimum, 541 feet. Greatest inclination 40 feet per mile; constructed in 1832; cost $90,000.

HARRISBURG, PORTSMOUTH, MOUNTJOY AND LANCASTER

PENNSYLVANIA. 131

RAIL-ROAD. This company was incorporated in 1832; since which time various supplements to their charter have been enacted by the legislature, under which their corporate privileges are now exercised. The road was commenced in 1836, and a single track completed in September, 1838. It is proposed to lay down, immediately, another track of the best T rail iron. Length, 35.50 miles. Maximum inclination 42.24 feet per mile, but generally under 35 feet; the radii of the curves are mostly 2640 feet, and only one where the radius is less than 1000 feet. More than 22 miles of the road is perfectly straight. It has one tunnel, 850 feet in length. Total cost of the road, locomotive engines, &c. $859,537 03 or $24,212 31 a mile.

The gross receipts of the company for tolls were, in 1837, $38,536 44; in 1838, $64,532 94, and in 1839, $92,894 72. Total in those three years, $195,964 10; from which deduct expenses, $139,524 50; net gain $56,439 60.

Two dividends, amounting to 10 per cent. on the cost, were declared in 1839: leaving a surplus of $16,614 72, which has been partly applied to the liquidation of the company's debts, and partly expended on the repairs of the rail-road. Already the sum of $49,146 43 has been expended in repairs, which became necessary in consequence of the settlement of the embankments and the abrasion of the slopes. This road forms a section of the line towards Pittsburg, commencing on the Columbia and Philadelphia Rail-road in Lancaster, and terminates at Harrisburg, where the Cumberland Valley Rail-road commences.

CUMBERLAND VALLEY RAIL-ROAD, is a continuation of the Lancaster and Harrisburg Rail-road. After leaving Harrisburg and crossing the Susquehanna, the road proceeds with a general ascending grade, nearly due west, to Carlisle, in Cumberland county; thence, gradually curving towards the southwest, it enters Newville. Here the road taking a south-west direction, and passing through Shippensburg and Green Village whence the grade descends, enters Chambersburg in Franklin county. Length 50 miles. This road constitutes a part of the great western route to Pittsburg, &c.

FRANKLIN RAIL-ROAD, is a prolongation of the preceding. Its course, after leaving Chambersburg, is nearly south, which is pursued, and passing through Greencastle, and along the left bank of Conecocheague creek, enters and terminates at Williamsport on the Potomac, in the state of Maryland, where the road intersects the Ohio and Chesapeake Canal. Length 30 miles.

CHAMBERSBURG AND PITTSBURG RAIL-ROAD. The line proposed for this work, commences at Chambersburg, passes through Cumberland valley, crossing the west branch of Conecocheague creek some distance below Loudon, thence ascending the side of Cove Mountain, it reaches Cowan's Gap, and descending the valley of Augwick creek, by the Burnt Cabins, to Sidling Hill Run, it proceeds up that stream. Passing through Well's valley, the line meets Ray's Hill, or properly a point where the Harbor Mountain joins the Broad Top Mountain, where a tunnel will be necessary. The line then proceeds, and, passing into Woodcock valley, and through the valley of Bloody Run, enters the town of Bedford. Thence over Dry Ridge and Deeter's Run, it encounters the main ridge of the Allegany, which must be tunnelled; and then passing on three miles north of Somerset, it ascends Laurel Hill; thence to Laughlintown, and along the valley of Loyalhanna, through a gap in Chesnut Ridge, it reaches Greensburg. From Greensburg it proceeds along Brush and Turtle creeks, and thus gains the valley of the Monongahela, which is followed to Pittsburg. Length of line 243 miles. It attains an elevation of 2081.69 feet at Laurel Hill.

Other lines from Chambersburg to Pittsburg have also been surveyed by the state engineers; but as no route has yet been definitively located, any further notice of them at this time, is deemed unnecessary.

YORK AND WRIGHTSVILLE RAIL-ROAD. This road, although the work of a private company, may be regarded as an extension of the Columbia and Philadelphia Rail-road. It unites with that road at Columbia, crosses the Susquehanna to Wrightsville, and thence proceeds to York, where it meets the unfinished rail-road to Gettysburg, and the Baltimore and Susquehanna Rail-road, now in operation. Length 13 miles.

STRASBURG RAIL-ROAD, in Franklin county, extends from the Cumberland Valley Rail-road to the town of Strasburg; 7 miles in length.

PENNSYLVANIA. 133

MARIETTA RAIL-ROAD, from Columbia to Marietta.

BALTIMORE AND YORK RAIL-ROAD. See Susquehanna Railroad, Maryland.

PHILADELPHIA AND READING RAIL-ROAD. This road has its point of outset at the foot of the inclined plane, on the Columbia and Philadelphia Rail-road, on the west side of the Schuylkill, about three miles from the city of Philadelphia. Ascending the right bank of the river Schuylkill, through Montgomery, Chester and Berks counties, the road enters the town of Reading in the last named county. It was commenced in 1835, under the orders of a joint stock company, and opened for public use on the 17th of July, 1838. A branch leaves the main line at the Falls of the Schuylkill, and thence proceeds in an eastwardly direction, and intersects the west bank of the Delaware, at the village of Richmond, three miles from Philadelphia. This branch is designed for the accommodation of the coal business. The entire length of this road from the Delaware to Reading, is 59 miles : and from its junction with the Columbia Rail-road, 54 miles. An extension of this work to Pottsville, 36 miles from Reading, is nearly completed : its structure is similar to that of the southern section. The heaviest grade from Reading to Philadelphia is 19 feet to the mile, for about 17,700 feet; between those points, there are 152,600 feet of level, and the remaining distance is divided into grades, varying from 1.5 to 11.8 feet per mile. All of which, as well as those between Reading and Pottsville, descend in approaching Philadelphia. About 30 miles above Philadelphia, the line passes through the Black Rock Tunnel, which cost $150,000. At Flat Rock, 8 miles from Philadelphia, is another tunnel, 960 feet long, cut through gneiss rock. Near Port Clinton is a third tunnel, 1600 feet long. The shortest radius of curvature is 819 feet, and but 1480 feet struck with this radius. The other curves generally average from 2000 to 3000 feet radius. Total cost, including Pottsville extension, $5,000,000.

Plan of construction.—The H rail is employed, weighing $45\frac{1}{8}$ lbs. per yard lineal ; each bar is $18\frac{3}{4}$ feet in length, with square ends, and weighs, on an average, 282 lbs., or 8 bars to the ton. With exception of the square ends, the form of the rail resembles that on the Washington branch of the Baltimore and Ohio Rail-road, except that it is 5⅛ lbs to the yard heavier than the latter.

The rail is laid upon the white oak sleepers, or cross ties, 7 feet in length, and hewn upon the upper and lower sides, so as to have a flat surface for the under bearing, and a similar one for the rail to rest upon of 8 inches wide; the depth of the sleeper being 7 inches uniformly. These are laid 3 feet 1½ inches apart from centre to centre, and cost, upon an average, delivered at distances apart of about two miles, on the graded surface of the road, about 60 cents each. Timber is scarce and dear upon the Schuylkill, and it was said that these were brought by the Union Canal from Huntingdon county. Each sleeper is laid upon a prism of broken stone, deposited in a trench 14 inches deep, 12 inches wide, and 9 feet long, transversely of the line of the track. The cost of broken stone was, on an average, (for this the first track) $1 10 per perch of 25 cubic feet, delivered in heaps 10 feet apart on the road surface. Two sizes of broken stone were used, the one to pass through a two inch, the other through a 3 inch ring, the larger of which constitute the lower portion of the mass. The stone were placed and compacted in three different layers, one upon the other. The spaces between the sleepers are filled with clay, or any material most convenient to be obtained. This filling reaches the top surface of the sleepers in the middle of the track.

Every sleeper (except where there is a chair) is notched to a depth of about one-fourth of an inch, to receive the lower web of the rails. These notches cost 5 cents per sleeper, which is not included in the 60 cents above-mentioned.

Of the *fastenings*, it may be observed, that the rails, at their joinings, rest upon cast iron chairs, let into the sleepers by means of notches cut for that purpose. The chair is 6 inches square at its lower surface, where it is five-eighths of an inch in thickness. Upon that side of the chair situated upon the outer side of the track, and upon the entire length of the chair, there is a portion of the casting having an upward projection, and passing over the lower web of the rail upon that side, and thence to the stem of the rail; and also extending to, or very nearly to, a contact with the under side of the *upper* web. Through this upper projecting part of the chair, there are two square countersunk holes, to receive square bolts, with heads formed to fill the said countersunk holes: each bolt passes through one of these holes in the chair horizontally, and likewise through a hole in the stem of the rail, near its end. The hole in the rail, however, is not precisely square, as it is in the chair, but is three-fourths by seven-eighths of an inch, and situated at a clear distance of three-fourths of an inch from the end of the rail. The hole in the chair is for a bolt five-eighths square, and the head of the bolt to fill the countersink, is fifteen sixteenths. Upon the inner side of the rails, a nut screws upon each bolt, to hold the ends of the two rails to the chair, and in proper line, whilst the hole in the rail is wider than the bolt to allow for contraction and expansion from change of temperature. The bolt and nut weigh seven ounces, and the chair $10\frac{1}{2}$ lbs., and is held in place by means of four spikes, the heads of which pass over the edge of the chair, whilst their stems are driven into the sleepers, and also fill recesses left for that purpose in the corners of the chairs in casting them. The same kind and size of spike is used to fasten the rail to each sleeper, (except where the chairs are) the head of the spike passing over the edge of the lower web on each side of the rail. The spikes are six inches in length, and their stems are three-fourths by five-eighths of an inch, and they weigh about three-fourths of a lb. each. It is thought that the stem should be square, and the length $4\frac{1}{2}$, or at most 5 inches.

The varied cost of the iron rails at Philadelphia, averaged about $60 per ton. And the cost of the conveyance to the road, by means of the Schuylkill Navigation, was $2 60 per ton.

There are in the mile of track,

Bars of rails, in number 563, weighing			.	71 tons.
Chairs,	. do.	563,	do.	5,910 lbs.
Spikes,	. do.	7,882,	do.	4,524 "
Screw bolts & nuts, do.		1,126,	do.	481 "
Sleepers of wood, do.		1,689.		

The track cost an average rate of $1 50 per sleeper, or $2,533 per mile, exclusive of the cost of all the iron materials, at Philadelphia.

The cost of *laying down* this single track of rail-way, consisting of excavating the trenches to receive the broken stone—putting down the broken stone—laying, notching, and adjusting the sleepers—putting on the chairs and the iron rails complete—*has been, on an average,* 40 *cents per sleeper,* or $675 60 per mile of track : to which add, for contingencies, such as cutting the iron bars, in order to make the joinings of each two have a position opposite to the middle of the length of the opposite rail, or bar, (this being a condition uniformly observed in the track) extra transportation, cleaning the side ditches, making crossings, &c. &c., say about $200 per mile.

The above-mentioned 40 cents per sleeper, or $675 60 per mile, is included in the aforesaid $1 50 per sleeper, or $2,533 per mile. The contracts for laying down the rail-way were made at so much per sleeper, viz. 40 cents as above.

The entire cost of the single track, as laid, is stated to be $7,617 per mile, inclusive of materials and workmanship.

The Tunnel, about 30 miles from Philadelphia, is 1932 feet in length, 19 feet wide, and $17\frac{3}{10}$ feet in extreme height. The sides are cut perpendicular at a height 10.9 feet from the bottom of the grade line, which is ten inches below the top of the rails. Above this the form of the cross section is that of a semi-ellipse, rising 6.4 feet. The faces of the excavations of the openings at the ends of the tunnel, are respectively 47 and 55 feet in height, and these are secured by well dressed masonry. Except at the ends, no masonry is required, as the rock, called here *Grauwacke slate,* appears to be of sufficient tenacity, to justify dispensing with arch masonry.

Stone Viaduct.—At the northern end of the tunnel, the road immediately crosses the Schuylkill to its left bank, by means of a stone bridge.

Length 4 spans, each 72 feet, . . 288 feet.
3 piers, " 8 " . . . 24 "

with circular wing walls to support the banks, &c. Width of structure, from out to out, 18 feet 4 inches. Roadway above low water mark 24 feet. Versed sine of each arch $16\frac{1}{2}$ feet, the form being that of a circular segment of $47\frac{1}{2}$ feet radius. The abutments and piers are founded upon the rock from 8 to 12 feet below the surface of low water, by the use of coffer dams. The work is laid in Roman cement below the water surface, and in common mortar above that line. The whole exterior is of cut stone, and has a very light and beautiful appearance.

PENNSYLVANIA. 137

The Philadelphia and Reading Rail-road, with its extension to Pottsville, completes the line of communication between Philadelphia and the anthracite coal region of Schuylkill county: and comes in direct competition with the Schuylkill Navigation, which unites the same points, and pursues nearly the same course as the former. It connects with the Mount Carbon Rail-road, and by means of it, with the Danville and Pottsville Rail-road. It will also, at the same place or at Mount Carbon, half a mile below Pottsville, be connected with the Mill Creek and Schuylkill Valley Rail-roads, either by the extension of one or other of the last named rail-roads, or by a branch of the Reading Rail-road of less than two miles to Port Carbon. At Schuylkill Haven it connects with the Mine Hill and Schuylkill Haven Rail-road; and at Port Clinton, 15 miles below Pottsville, it connects with the Little Schuylkill Railroad, by which the Lehigh Coal and Navigation Company can transport their coal to the Reading Rail-road. It will thus be perceived, that the Reading Rail-road unites with all the rail-roads in the coal region of Schuylkill county, by which the coal is, at present, brought in cars to the Schuylkill Canal.

Some idea of the character and construction of the Philadelphia and Reading Rail-road may be formed from the fact, that an engine weighing 11 tons, conveyed over the road from Reading to Philadelphia, 101 cars, with a gross weight, including the engine, of 423 tons, at an average speed of 10 miles per hour. Among the freight were 2002 barrels of flour, weighing $190\frac{1}{2}$ tons! Amount of freight for this trip, $835 19, and expenses of every sort, including the return of the empty cars, &c., $105 94; net profits, $729 23.

LITTLE SCHUYLKILL RAIL-ROAD. Commences at Port Clinton, at the junction of the two principal branches of the Schuylkill; and extends to the mines at Tamaqua, at the foot of the Broad Mountain. Length 23 miles; single track, although graded for a double track. The radii of curvature are in general from 477 to 1000 feet. Rise, $406\frac{7}{12}$ feet.

DANVILLE AND POTTSVILLE RAIL-ROAD. Commences at a point on the Mount Carbon Rail-road, $2\frac{1}{2}$ miles N. W. of Pottsville; by a deep cut and tunnel, of 700 feet in length, the road passes into and along the valley of Mill Creek, until it reaches, by four inclined planes, the first summit on the Broad Mountain, 1014 feet above Sunbury. Thence it descends the Broad Mountain by a plane, which depresses the road nearly 400 feet to a level, $2\frac{1}{2}$ miles in extent, when the sixth inclined plane conducts it to another level of four miles. The line then proceeds and gains the summit between the Mahonoy and Shamokin creeks, by the seventh plane and an ascending grade, and then descends the Shamokin valley to Sunbury on the Susquehanna; length 44.54 miles. A branch, 7 miles long, from the main line to Danville is proposed; entire length 51.54 miles.

The eastern section of this road was opened for use on the 24th of September, 1834. On this section there are 1 large, and 4 inferior inclined planes. The former, in Mahonoy valley, is 1650 feet in length, and overcomes an elevation of 345 feet, and the 4 latter 700 feet.

The coal tunnel on the Girard estate, which opens a communication between the Mahonoy and Shenandoah valleys, is 2500 feet long.

LITTLE SCHUYLKILL AND SUSQUEHANNA OR CATAWISSA RAIL-ROAD, is a prolongation of the Little Schuylkill Rail-road, though owned by a different company. The Company was incorporated by the legislature of Pennsylvania, in the year 1830, with a capital of $300,000, for the purpose of extending the Little Schuylkill Rail-road to the town of Catawissa, on the *North Branch* of the Susquehanna. By subsequent legislative enactments, however, the capital stock of the Company was increased to *two millions* of dollars; and the managers were authorized to extend their road to Williamsport, on the *West Branch* of the Susquehanna. They are also permitted to hold *five thousand acres* of coal land; to exercise mining privileges; and to make whatever contracts they may consider necessary, with other rail-road and canal companies, for the transportation of their coal to market.

This road commences on the northern termination of the Little Schuylkill Rail-road, and proceeding northward, through Lindner's Gap, and a tunnel, 1150 feet in length, in the ridge which divides the waters of the Schuylkill from those of Catawissa creek, it enters and pursues the valley of the Catawissa, to the village of that name, on the North Branch of the Susquehanna. Here it intersects the line of the Pennsylvania Canal, which will extend to the New York boundary. From Catawissa the road ascends the left bank of the Susquehanna a short distance, then crosses that river to Bloomsburg, where it enters the valley of Little Fishing Creek, which it ascends, passing through the village of Millville to Cox's Gap, where it traverses a summit 479 feet above Catawissa. Descending the northern declivity of the Muncy Hills to Muncyboro, and thence with the left bank of the West Branch of the Susquehanna, the line is conducted to Williamports in Lycoming county, where it unites with the Williamsport and Elmira Rail-road.

A branch line, 12 miles in length, with a descending grade, leaves the main road near Lindner's Gap, about one mile southeast of the summit, traverses the valley of the Quakake, and intersects the Beaver Meadow Rail-road, about four miles from the Lehigh, and below its inclined planes. At Lindner's Gap is a plane 1900 feet long, whose angle of inclination is $4° 58'$, overcoming an elevation of 165 feet. In the section from Catawissa to the summit tunnel there is no grade exceeding 33 feet per mile: from that point to Tamaqua the maximum inclination is 66 feet per mile, and the total descent 740 feet. There are two tunnels on the main line, and one inclined plane on the Beaver Meadow branch road. Total cost, as estimated by the principal engineer, exclusive of engineering, $1,622,117. Length of main line 106 miles; general course north-west. Though it appears by the charters of the two companies, that the Little Schuylkill company's works should terminate 9 miles north of Tamaqua, we have considered them as terminating at Tamaqua, and framed our descriptions accordingly.

WILLIAMSPORT AND ELMIRA RAIL-ROAD. This road leaves Williamsport at the termination of the Tamaqua road, and pursues the left bank of Lycoming creek; passing through the village of Ralston and the county of Bradford, it enters Chemung county, in New York, and terminates at the town of Elmira, the southern terminus of the Chemung Canal. Length 73.50 miles. General course, N. N. E.

CORNING AND BLOSSBURG RAIL-ROAD, partly finished and the remainder in progress, from Blossburg, in Tioga county, Pa., to Corning at the western termination of the Chemung Canal, in Steuben county, N. Y. Length 40 miles. With the addition of about fifteen miles of rail-road, which are proposed, from

Blossburg to the Williamsport and Elmira Rail-road, an unbroken chain of improvement by canal and rail-road, from Philadelphia to Buffalo and the Falls of Niagara, will be established. This is a most important improvement as it affords a new and convenient route to the central and western parts of New York, and is the first opening from those parts to the coal region of Pennsylvania. The chartered name of the company under whose direction this work was commenced, is the " Tioga Navigation Company."

CATAWISSA AND TOWANDA RAIL-ROAD, as proposed, will diverge from the Little Schuylkill and Susquehanna Rail-road at or near Millville, and pass up the east bank of the Little Fishing Creek, thence by the head waters of the Muncy and Loyalsock, through Towanda and Athens, to the New York state line, where it will intersect the New York and Erie Rail-road, and form a connection with the rail-road between that point and Ithaca, at the head of Cayuga Lake.

SUNBURY AND ERIE RAIL ROAD. The necessity for a continuous rail-road communication from Philadelphia to the great lakes, was long since suggested and is now universally admitted. In consequence of the increase of trade between eastern and western Pennsylvania, it has become an object of the utmost importance to improve those channels of communication upon which the continued increase of that trade depends. Influenced by this consideration, some of the citizens of Philadelphia, in conjunction with others, concerted measures for the construction of a rail-road to extend from the town of Sunbury to that of Erie on the southern shore of Lake Erie, and thus complete the line from Philadelphia to that point: a charter having been obtained in 1837, the surveys which soon followed resulted in the adoption of the following line :—Commencing at the town of Erie, it passes in a south-east direction, and ascends the escarpment which separates the waters of Lake Erie from those of French Creek. Having attained the summit, the line enters the ravine of Boeuf Creek, and thence crossing by an eastern course the north branch of French Creek, it surmounts the ridge between the head streams of that creek and those of the Broken Straw Creek, whose valley is pursued to the mouth of Kenjua Creek, in Warren county. Here the line deflects towards the south-east, and ascending the valley of the Kenjua, proceeds over the high table-land of

M'Kean county, and falls into the ravine of the Driftwood branch of the Sinnemahoning, whose valley is then entered and followed to its junction with the West Branch of the Susquehanna in Clinton county. From the confluence of the Sinnemahoning, and West Branch, where it meets the western terminus of the state canal, the line proceeds along the right or south bank of the West Branch, and terminates at a point opposite to Dunnstown. The route thence to Sunbury has not been definitively located, nor is its point of connection with existing works yet determined. Owing to the deranged state of the currency, nothing further than a survey and location of the line has yet been done in its execution. By means of this rail-road, and either the Danville and Pottsville, or the Susquehanna and Little Schuylkill Rail-road, and the works now completed or nearly so, between Port Clinton and Philadelphia: a continuous railroad communication will be opened from the latter city to the town of Erie, a distance of about 420 miles.

The great importance of such a communication, and the deep interest which the friends of internal improvement take in its successful completion, are abundantly exemplified in the enthusiasm which animated the members of the late convention. The merchants of the east and the manufacturers and farmers of the west, cannot fail to perceive how intimately their future success in trade is identified with the issue of the great effort now making to open the way to a more intimate connection between them, and thus to promote their mutual prosperity. At present there is but little intercourse between the inhabitants of the east and those of north-western Pennsylvania, for the obvious reason that the necessary facilities of a direct communication are wanting. The citizens of Pennsylvania, therefore, could not but view this enterprise with peculiar favour, as tending to unite more closely the interests of every member of the great Pennsylvania family, and thus to cement that bond of union, which should form one of the leading objects of all such efforts.

It is proposed to extend a branch from the Sunbury and Erie Rail-road to Pittsburg. Surveys to a limited extent have been made, chiefly to ascertain the practicability of crossing the main ridge of the Alleganies. These surveys prove satisfactorily that a rail-road, without inclined planes, may be constructed from Lock Haven, on the Sunbury and Erie Rail-road, along the valley of Bald Eagle Creek, through Emigh's Gap of the Allegany mountain, into the Kiskiminitas valley ; whence, to Pittsburg, no uncommon difficulty is likely to interpose to prevent the accomplishment of this important connection.

MOUNT CARBON RAIL-ROAD, commences at the lower landings of Mount Carbon, passes through Pottsville, and thence up the valley of Norwegian creek, where it unites with the Danville and Pottsville Rail-road. Length, including two branches, 7.24 miles. Cost $118,000. Rise 246.50 feet.

SCHUYLKILL VALLEY RAIL-ROAD, extends from Port Carbon, where the Schuylkill Navigation terminates, and follows the valley of the Schuylkill to Tuscarora. Length 10 miles. It has 20 branches extending from this road in various directions, the aggregate length of which is 15 miles. Cost of main line, $5,500 per mile. There are two sets of tracks, the width of each is $3\frac{4}{12}$ feet. The curves are numerous and many of them abrupt.

SCHUYLKILL RAIL-ROAD, 13 miles in length ; cost, $7,000 per mile.

MILL CREEK RAIL-ROAD, from Port Carbon to the coal mines near Mill Creek. Length of main line 4, and of branches 5 miles ; cost $20,000.

MINE HILL AND SCHUYLKILL HAVEN RAIL-ROAD, commences at Schuylkill Haven, extends along the west branch of Schuylkill, through Mine Hill Gap, and terminates at the coal mines in that vicinity. Length, including two branches, 20 miles ; cost $181,615. The curves have radii from 400 to 500 feet ; maximum grade, 30 feet per mile ; 2 inclined planes ; 50 bridges ; 12 culverts ; highest embankment 21 feet.

MAUCH CHUNK RAIL-ROAD, extends from Mauch Chunk to to the coal mines, 9 miles, exclusive of branches which are nearly 16 miles in length ; constructed in 1827 ; elevation of the mines above the Lehigh, 936 feet. The curves are generally abrupt, most of the radii being only 190 feet. Some of these have been improved. The line ascends, in some parts, at the rate of 133 feet per mile. Cost $3,500 per mile. Single track, with turn-outs, &c. ; width between the tracks 42 inches ; iron rails, one inch and three-quarters wide, three-eighths of an inch thick, fastened on wooden rails, sleepers mostly of wood.

ROOM RUN RAIL-ROAD, extends from Mauch Chunk to the

PENNSYLVANIA. 143

coal mines on Room Run. Length 5.26 miles. Rise 534.57 feet. Cost $76,111.

BEAVER MEADOW RAIL-ROAD, extends from Parryville on the Lehigh, 6 miles below Mauch Chunk, to the Beaver Meadow coal mines in Northampton county. Length 20 miles.

HAZELTON AND LEHIGH RAIL-ROAD, from the mines above Hazelton to a point on the Beaver Meadow Rail-road, 8 miles.

NESQUEHONING RAIL-ROAD, from the Nesquehoning coal mines in Northampton, to the Lehigh, 5 miles.

LEHIGH AND SUSQUEHANNA RAIL-ROAD, extending from White Haven on the Lehigh, to Wilkesbarre on the Susquehanna, forms a part of the Lehigh Coal and Navigation Company's works, and unites the Wyoming valley with that of the Lehigh.

It is 19.58 miles in length, with one tunnel and three inclined planes, by which it ascends the mountain from Wilkesbarre. From its northern terminus, the Wyoming Coal Company have constructed a branch rail-road, four miles in length, extending to their mines, and thence to the North Branch Canal. This road will be so constructed with iron T rails as to permit loaded boats to be conveyed over it and thus avoid the transhipment of their cargoes.

CARBONDALE AND HONESDALE RAIL-ROAD. This road extends from Honesdale, the western terminus of the Lackawaxen Canal, to Carbondale on the Lackawana river, in Luzerne county, Pa. It forms the concluding link in the chain of improvement from the Hudson river to the coal region of Luzerne county, Pa. Length, including branches, 17.67 miles; to which add about 4 miles of turn-outs and side lines; the entire length of single track is 21 miles. The road attains the summit of Moosic mountain, 912 feet above the mines, by seven inclined planes, worked by stationary power, and descends, 850 feet, by three self-acting planes. Minimum radius of curvature 1000 feet. About nine miles of this road consist of lofty truss work in place of embankments.

LYKENS VALLEY RAIL-ROAD, extends from the Broad Mountain through Bear Creek Gap, and thence on the north side of Berry's Mountain, to Millersburg on the Susquehanna, in Dauphin county. Length 16.50 miles.

Pine Grove Rail-road, from Pine Grove, in Schuylkill county, to the coal mines, 4 miles above.

Philadelphia and Trenton Rail-road, commences at Philadelphia, passes through or near the towns of Frankford, Bristol, Tullytown, &c. and terminates at Morrisville, opposite Trenton: length 26.25 miles.

Philadelphia, Germantown and Norristown Rail road, extends from Philadelphia to Norristown. About three miles from the former, the road to Germantown branches off and pursues a north course, whilst that to Norristown enters the valley of the Schuylkill, which it follows to Norristown, passing through Manayunk. Length from Philadelphia to Norristown, 17 miles. Entire length, including Germantown branch, 21 miles. The cost of the latter was nearly $50,000 per mile.

Philadelphia and Wilmington Rail-road, commences on the Southwark Rail-road at the intersection of Prime and Broad streets in Philadelphia, proceeds towards the southwest through Chester, in Delaware county, and terminates on and unites with the Wilmington and Susquehanna Rail-road, at Wilmington, in Delaware. Length 27 miles.

This road and the others which now form a continuous line of rail-road between Philadelphia and Baltimore, having been commenced by four several and distinct companies, it soon became obvious that an union of the various interests was indispensable to the full development of all the advantages, which, under judicious management, might be anticipated, from this important link in the great Atlantic chain. Measures were accordingly adopted, to re-organize the whole; and, after obtaining the sanction of the various legislatures, an arrangement was effected, on the 5th February, 1838, by representatives from the respective companies, and subsequently ratified by the stockholders, by which they were consolidated under the title of the Philadelphia, Wilmington and Baltimore Rail-road Company. In the construction of this work, several kinds of rails have been adopted for different sections of the route, in all of which, strength and consequent permanence have been made essential requisites. The bridge rail, weighing 40 lbs. per yard; the T rail, weighing 56 lbs. per yard; and the heavy bar rail, 1 3-4 inches in thickness, by 2 1-2 inches in breadth, weighing 40 lbs. per yard, are used throughout the whole, with the exception of a portion of the route between Philadelphia and Wilmington, upon which the heavy plate bar has been laid.

The superstructure of the road consists of longitudinal sills, connected by cross ties of locust, red cedar, or seasoned white oak, and surmounted by longitudinal string pieces of Carolina heart pine, on which is laid the iron rail. Upon the greater part of the road, however, the strength of the iron bar is such, as to render unnecessary the use of the longitudinal string piece, the bar being supported by the cross tie alone.

Between the city of Wilmington and the Susquehanna river, (see Maryland,) the roadway is graded thirty-five feet in width with superior bridging, all but one being built of the most substantial stone masonry and brick arches, making them secure from risk of fire. Upon other portions of the road, the surface width is twenty-five feet, having, throughout nearly the whole length of the line, a surface graded sufficiently wide for two tracks of railway. The whole distance was contracted for and finished by different contractors, in various quantities of from five to ten miles in extent, and amounts from $10,000 to $60,000.

The total receipts of the road for the year 1839, were $490,635 55, exceeding those of the preceding year by $118,720 61, and the whole number of persons conveyed on the road during the year 1839, was 213,650, a great portion of whom were through passengers. The receipts from passengers amounted to $416,974 76, and from the transportation of merchandise, $39,239 27, and for the United States mail $27,497. The dividends for the same year, were seven per cent. on $4,379,225 17, the total amount expended by the company for road construction, building, travelling apparatus, &c.

GETTYSBURG RAIL-ROAD. This road, the construction of which was commenced under the authority of the state government some years since, extends from York in York county, in a south-west direction, through Gettysburg, in Adams county, to the village of Clear Spring, on the Potomac, in Maryland. Length, about 91 miles, 13 of which are in the state of Maryland. After expending $622,891 61, and incurring liabilities to the amount of $145,307 78, in addition, the legislature, by a resolution passed February 19th, 1839, directed the suspension of the work from and after the 1st of March ensuing, with the implied understanding that it was not to be resumed. That portion of the route which extends from York to Gettysburg, about 28 miles in length, passes through one of the most populous sections of the state, and if completed, would form an important link in the chain of rail-roads to Pittsburg. It is much more direct than the Harrisburg route, as will appear on consulting the map, and in this point of view, it should have received the earliest attention. When the great amount ($768,127 39) already expended on the work, and the probable benefits which would result from the completion of the former section, are considered, its entire abandonment in its present advanced stage, is matter of regret to the friends of internal improvements. From the large amount of money and time consumed upon this portion of the line, it may be inferred that the period of its completion, was not very far distant when it was determined to suspend the work. Unless the expense of construction should greatly exceed the average cost of other American rail-roads, a small additional appropriation on the part of the legislature would have ensured the completion of the York and Gettysburg division. With regard to the question of an abandonment of the work altogether, the inquiry should have been, not whether the road would yield the current interest on the whole investment, but whether the revenue to be derived from the section under consideration, would justify such an additional expenditure as would effect its completion. That it would produce a handsome interest on this *additional* amount, no one who is familiar with the country, can entertain a doubt. In the abandonment of the remaining division, from Gettysburg to the Potomac, the legislature is fully justified by the almost unparalled cost which would have attended its construction, and the certainty of an inadequate return. Nearly the whole line from Gettysburg to Clear Spring, as located, would have required a continued succession of tunnels, bridges, culverts and embankments, forming altogether one of the most irregular profiles we have yet had occasion to examine.

Aggregate length of Canals in Pennsylvania, 974.06 miles.
Aggregate length of Rail-roads in Pennsylvania, 953.58 miles.

DELAWARE.

RAIL-ROADS.

PHILADELPHIA AND WILMINGTON RAIL-ROAD, see Pennsylvania.

WILMINGTON AND SUSQUEHANNA RAIL-ROAD, see Maryland.

NEW CASTLE AND FRENCHTOWN RAIL-ROAD, extends from New Castle on the Delaware, 35 miles below Philadelphia, to Frenchtown, on Elk River, one of the head streams of Chesapeake Bay. Length 16.19 miles; radius of the largest curve 20,000 feet; of the least, 10,560 feet. The inclinations (with one exception of 29 feet a mile,) vary from $10\frac{1}{2}$ feet to $16\frac{1}{3}$ feet a mile; 4 viaducts, 29 culverts; commenced in 1830, completed in 1832; entire cost of road, apparatus, &c. $400,000.

The first track of this road, of $16\frac{1}{2}$ miles long, was laid with the plate rail upon longitudinal strings of wood, resting, as usual, upon notched sleepers of wood.

The rail-way of the second track, which has been in use four years, is laid with an H rail, very similar to that of the Camden and Amboy rail-road. The mode of attachment of the bars to each other, at the joinings, is like that upon the road just mentioned. The rail rests on flatted sleepers, three feet apart from centre to centre, reposing on three inch plank as an under sill: the last rests on the natural material forming the road bed. The rail is fastened down with spikes of the brad form. Under the rails, at their joinings, is introduced a plain plate of wrought iron, equal in width to the bar of the rail and about five inches long. The ends of the rails are square, and the bars fifteen feet long.

This road, in connection with the steamboats on the Delaware and Chesapeake, forms a part of one of the principal routes between Philadelphia and the south.

NEW CASTLE AND WILMINGTON RAIL-ROAD, contemplated, was incorporated in January, 1839. The object of this road is

148 DELAWARE.

to open a communication for the transit of merchandize from New Castle to Philadelphia, when the navigation of the Delaware is obstructed by ice. Length, 5 miles; probable cost, about $80,660.

DELAWARE RAIL-ROAD. The line of this proposed work begins on the Wilmington and Susquehanna Rail-road at Wilmington, extends through the entire length of the state, and terminates on Nanticoke Creek, in the vicinity of Seaford. Though great inducements have been offered to enlist the co-operation of capitalists in this enterprise, no effectual step has yet been made by the legislature to insure its execution at present.

CANAL.

CHESAPEAKE AND DELAWARE CANAL, commences at Delaware city on the Delaware, about 42 miles below Philadelphia, passes through St. George's meadows and along the ravine of Broad Creek, and thence into that of Back Creek, a tributary of Elk river, which falls into Chesapeake Bay; length 13.63 miles; 66 feet wide at top water line; 10 feet deep; 2 lift and 2 tide locks, 100 by 22 feet in the chamber; completed in 1829; cost $2,750,000.

MARYLAND.

RAIL-ROADS.

BALTIMORE AND OHIO RAIL-ROAD, is the work of a joint stock company, incorporated on the 28th of February, 1827. The execution of the work commenced on the 4th of July, 1828, and is now gradually proceeding towards the Ohio river, agreeably to the original design. It commences at the depot of the company in Baltimore, passes in a south-west direction to Elkridge landing, and thence along the valley of the Patapsco, to Parr's spring, thence into and with the ravine of Bush Creek, which it follows to the Monocacy, which is crossed about 3½ miles from Frederick, thence along the Monocacy valley to the Point of Rocks on the Potomac, and thence to Harper's ferry, 80.50 miles from Baltimore; commenced in 1828; capital $5,000,000. A road of a single track extends from the main line after crossing the Monocacy, to Frederick, 3½ miles.

The road bed is 26 feet wide. Maximum grade on the first 29 miles, 22 feet per mile; on the next 11 miles, the grades vary from 22 to 30 feet; on the next 4 miles, from 30 to 47½ feet. First inclined plane, 41 miles from Baltimore, ascends $80\frac{375}{1000}$ feet in a distance of 2,150 feet. The second plane ascends $99\frac{605}{2000}$ feet in 3000 feet. The summit, at Parr's Spring Ridge, is 813¼ feet above mid-tide. The line then descends by an inclined plane, 3200 feet long, descending 159.63 feet, and another, 1900 feet long, and descends 81.350 feet. All the planes are straight. On the next five miles, the grade does not exceed 37 feet, except at two places, extending 1176 feet, where it is 52 feet to the mile; thence to the Point of Rocks it is more level. Of the line from Baltimore to the Potomac, 33.12 miles are straight; 13 miles and 3968 feet are curved with radii not less than 955 feet; and about 21 miles varying from 395 to 955 feet; one curve, 1400 feet long, has a radius of 318; another, 1100 feet long, has a radius of 337 feet.

On the branch to Frederick, 2 miles and 1034 feet are composed of straight lines; the minimum radius of curvature of the remainder, is 477 feet. The maximum grade per mile 30.096 feet.

The viaducts are, with two exceptions, of stone; some of them are very splendid and costly. A cut, near Baltimore, is 70 feet deep, the mere excavation of which, and removal of the earth cost $122,118; and an embankment across Gadsby's run has a maximum elevation of 57 feet. The whole work indeed, from Baltimore to Ellicott's, has been executed at an unusual expense. There are thirty-three viaducts between Baltimore and the Potomac.

The superstructure is various; it consists of stone sills, stone blocks and wooden sleepers, on different parts of the line. Forty miles of single track are composed of granite sills, 8 inches thick, 15 wide, and of various lengths: these are laid in trenches, filled with broken stone. Six miles of single track are composed of stone blocks and wooden string-pieces, 6 inches square. The blocks are 4 feet apart, from centre to centre. The line hence to the Potomac, rests on wooden sleepers, four feet apart and imbedded in broken stone. The sleepers are hewn out in the centre to make room for the horse-path. The iron bars are 15 feet in length (pierced by 11 oblong holes) $2\frac{1}{2}$ inches wide, $\frac{5}{8}$ inch thick, and bevelled at the ends.

The extension of the Baltimore and Ohio Rail-road to the Ohio river, has been located, and a part of the road is now in progress. The line on leaving Harper's ferry, to which point the road is completed, ascends the west bank of the Potomac, to Opequan Creek, where it turns towards the south-west, and, following the valley of that creek, enters Martinsburg, in Berkeley county. Thence by a nearly direct north-west course the line is conducted over the ridges of Berkeley and Morgan counties, and crosses the Potomac into Maryland. After crossing the Potomac, it turns abruptly and pursues a south-west course along the north declivity of the Potomac to the mouth of Town Creek in Allegany county; and thence curving towards the north-west, proceeds, by the river bank, to the town of Cumberland. Here the road leaves the Potomac, and at a distance of 7 miles, passes into Pennsylvania, and descends the valley of Casselman's river, whose southern bank is followed to its dis-

MARYLAND. 151

charge into the Youghiogeny, thence through gaps in Sugarloaf mountain and Laurel Hill, in Fayette county, and running near Uniontown, it enters and pursues the valley of Redstone Creek to Brownsville, on the Monongahela. From Brownsville its course is nearly direct, through Washington county, until it reaches the western boundary of Pennsylvania and re-enters Virginia, when it descends the valley of Wheeling Creek, and finally terminates at the town of Wheeling, on the Ohio river. The entire length of the line from Harper's ferry to Wheeling is about 200 miles; and 280.50 from Baltimore.

BALTIMORE AND PORT DEPOSITE RAIL-ROAD, commences on the line of the Baltimore and Ohio Rail-road, at the intersection of President and Fleet streets, and extends, through Canton, to Havre de Grace on the Susquehanna, in Hartford county. Length 36 miles. This road forms a part of the rail-road line to Philadelphia, about 95 miles in length. Maximum inclination 20 feet per mile; minimum radius of curvature, 2000 feet, with the exception of a single curve of 1273 feet radius, at its entrance into Havre de Grace.

Plan of construction.—It was graded to a width from 18 to 22 feet, with the view of gradually increasing the breadth of the road bed in the future course of repairs. The railway structure employed consists of a sill, under each line of rails, of sawed white pine, 6×8 inches in the section, and of various lengths, from 12 to 40 feet. These sills are laid on their flat sides, in longitudinal trenches of a width and depth equal to the section of the sills, whose upper surfaces are therefore in the plane of the graded surface of the road. Upon the sills are placed at uniform distances of three feet from centre to centre, cross ties of white oak and chesnut. These cross ties are eight feet in length, and of two sizes in the section, the larger being eight inches, and the smaller six inches diameter, clear of bark, at the small end, the larger and smaller sizes being placed alternately along the track. Each cross tie has four notches in it; two on the lower side, of a width of eight inches, equal to that of the greater dimension of the under sill, and two on the upper side, $7\frac{1}{2}$ inches wide in the middle, with a slant to accommodate the wooden key used in wedging fast the upper string piece. The thickness of wood left between the notches is invariably $2\frac{1}{2}$ inches. The lower notches embrace the under sills, which fit them accurately enough to prevent injurious lateral movement endwise of the cross tie when it is laid and adjusted, in doing which shallow cross trenches are dug to receive the rounded portion of the cross tie descending below the top of the sill. The cross ties received no other dressing than the notching to receive the sill and string piece. In the upper notches of the cross ties rest the string pieces, 6×6 in the section, of Norway or Carolina yellow pine. Upon a portion of the track a string piece, 5×6, was used to make up a deficiency in the supply of the quantity required of the larger scantling. The string pieces are laid in the manner usual in the railways in which they have been used in connexion with the plate rail. The rail placed upon the string piece is a bar weighing 40 lbs. per lineal yard, of a nearly rectangular section, $2\frac{1}{2}$ inches wide at bottom, $2\frac{1}{4}$ inches *full* wide at top, and $1\frac{3}{4}$ inches high. The lengths of the bars vary from 17 feet 9 inches to 18 feet 3 inches, their ends are cut off obliquely at an angle of 60° with the line of the rail. They are perforated vertically by 5 holes 11-16 of an inch in diameter, and of a circular section for $1\frac{1}{4}$ inches from the bottom of the rail upwards, the remainder of their depth, next to the top of the rail, being enlarged longitudinally of the rail, so as to form a countersink of half an inch deep and $1\frac{3}{8}$ inches long by full 11-16 wide. Two of the holes are one inch in the clear from the ends of the bar, and the intermediate three are at equal distances from each other, and from those at the ends, of about 4 feet 6 inches. The ends of the bars at their joinings are supported upon chairs or splicing plates of rolled iron, $5\frac{3}{4}$ inches long by $4\frac{1}{4}$ wide, and one-fourth of an inch thick. These plates have two small ledges or brads on the upper side, extending the entire length of the plate, parallel to each other, and a distance apart in the clear, equal to the breadth of the bottom of the rail which rests between them, and is prevented by them from moving to either side. Each plate has two holes in it, corresponding to those in the ends of the bars. Through these holes, and others in the same vertical line bored through the string piece, are passed bolts of about 9 inches long, with heads shaped so as to fill the countersinks in the upper part of the holes in the bar, and with threads upon their lower ends, upon which a nut is screwed up against the bottom of the string piece without any washer, thus holding the rail down upon the splicing plate and securing it from rising. The joinings of the bars are thus, by the bolt, and the ledges upon the splicing plate, kept in exact apposition. Through the intermediate holes in the bar are driven spikes, 6 inches long, and going $4\frac{1}{4}$ inches into the wood, with heads shaped to fill their countersinks, like those of the bolts. The heads of the bolts and spikes are left *full*, and are driven hard into the countersinks, so as to fill them up as accurately as possible, and afterwards dressed, or chipped off even with the top surface of the rail, to preserve its smoothness and continuity. The rail is placed in the middle of the string piece, and the joinings are made to fall between the cross ties, to allow of the screwing on of the nut at the bottom of the bolt. This is managed by some attention to selecting the bars with respect to their length, and in some cases by moving the cross tie along the track a sufficient distance, which can never exceed about half of its own breadth. No respect is paid to making the joints of the two lines of rails hold any fixed position with regard to each other.

It was intended to let the cross ties, (excepting the small part of them below the bed of the lower notch) together with the string pieces and the rails on them, stand entirely above the graded surface with which the top of the under-sill was designed to be coincident. The object of which arrangement was to lift the track above the reach of ordinary snows and mud, and to promote the preservation of the string piece by freeing it from contact with the ground, as well as to facilitate access to the bolts confining the ends of the rails, and render more easy the removal of all the timbers composing the track, as decay made their removal necessary. As this position of the track, however, deprived it of the support which is usually given to railway superstructures by the filling in and around them of earth or other heavy materials, the lower notch of the cross tie was adopted to connect it with the imbedded under sill, which, besides being of a size and weight capable of offering considerable resistance to motion, was held in its position by the surrounding ground. A few weeks' use of the road, however, with locomotives travelling with very high speeds, appears to have demonstrated the insufficiency of this precaution against the displacement of the superstructure of the track, which is now in the course of being filled with earth to the level of the top of the string piece. The track has been laid upon the soil naturally forming the surface of the road bed, which, for a considerable part of the length of the line, is of a material not retentive of water. Other parts of the road bed are, however, of a less favourable consistency, and will be affected by frost during the period of its action, though the derangement consequent thereupon will not, it is thought, be serious, as the weight and compact connection of the frame of the track will cause it to rise and subside with some uniformity of movement. Great attention has been paid to the promotion of effectual drainage in the cuts, by capacious ditches of a sufficient longitudinal slope. The cost of the above described railway has been per mile very nearly as follows:—

42,240 feet bd. measure, under sills, 6 × 8, at $13 32 per M. $562 50

1,760 cross ties, notched and delivered on road bed, at 32 cts. 565 20

35,000 feet bd. do., string pieces, 6 × 6, inclusive of wedges for fastening strings in cross ties, at an average of $18 40 per M. 644 00

586 splicing plates, weighing 1025 lbs., at 6 cts. 61 50

1,172 screw bolts for ends of rails, weighing 879 lbs., at 13 cts. 114 27

1,758 spikes for intermediate holes, weighing 1055 lbs., at 13 cts. 137 15

Transportation of materials by land and water, 175 00

Workmanship of laying track, . . 800 00

Cleaning out ditches, raising embankments and dressing road bed, ... 136 00

63 tons of iron rails, at $62 per ton, delivered in Baltimore, ... 3,906 00

$7099 62

Add for turn-outs, crossings and sidings, . 200 00

Do. superintendence, 3 per cent. . 212 98

Total cost of a single track, . . $7,512 00

BALTIMORE AND SUSQUEHANNA RAIL-ROAD, extends from

MARYLAND.

Baltimore, in Maryland, to York in Pennsylvania; 56 miles in length; summit 1000 feet above tide water; some of the curves are abrupt; the general structure is similar to the Baltimore and Ohio Rail-road.

It commences on Calvert street, Baltimore; proceeds up the valley of Jones Falls to Rowland's run, which is followed to its source, thence over the dividing ridge between Jones Falls and Gunpowder river, and thence by a nearly north course to York. The steepest ascent towards York, is 84 feet per mile, and descent 59 feet per mile. The least radius of curvature is 950 feet, with one exception of 820 feet. Graded for two tracks, only one is laid. Cost $16,185 89 per mile.

A branch diverges from the main line about six miles from Baltimore, proceeds along the valley of Jones Falls, and intersects the turnpike from Baltimore to Reisterstown, eight miles from its point of outset.

WASHINGTON BRANCH OF THE BALTIMORE AND OHIO RAILROAD, leaves the main line at the Patapsco river, about 8 miles from Baltimore, and proceeds in a general south-west direction, and terminates on Pennsylvania Avenue, in the city of Washington, distant 38.35 miles from Baltimore. Length 30.35 miles. The line is graded for two tracks, only one of which (excepting about 5.50 miles of second track, of the principal cuts) has been laid down. The highest grade is at the rate of 20 feet per mile, and the least radius of curvature, 1273 feet. It was opened for travel, August 25th, 1835.

Plan of construction.—An H rail of 40 lbs. to the yard, in bars of 15 feet in length, with scarfed ends at an angle of 60 degrees is employed. The rail is laid upon the middle part of a continuous string piece of wood, six inches square, resting upon, and keyed in the notches of sleepers, or cross ties of wood, 8 inches diameter at the smaller end, laid three feet apart from centre to centre; the latter resting upon continuous longitudinal sills, six inches square, imbedded in the road bed underneath the respective bearings of each line of rails. The several joinings of the bars of the railway, which happen promiscuously in any part of the length of the string piece, are secured, and the ends of the bars kept in their proper horizontal and vertical position by means of cast iron chairs. The chair is about six inches square, and weighs about 8 lbs. It has two downward projections, or flanges, grasping the two sides of the string piece to which it is fastened by a spike driven horizontally through a hole in each flange. The two ends of adjoining bars rest upon this chair, and they are secured on the inside of the track, both from sliding inwards, horizontally, and from rising up, by a *lip*, cast upon, and co-extensive with, that side of the chair, which lip extends upwards and laps over the lower web of the rails on that side ; whilst any sliding in the contrary direction, as well as a rising of the rail upon the outer side, is prevented by the square shank of a screw bolt passing up through a square hole in the chair, and in contact with the lower web of the rail, and by a plate of cast iron, weighing about 3 lbs. placed upon the top of the lower web of the two bars, and fastened down upon them by a nut on the top end of the said screw bolt, the screw part of which bolt passes up through a hole for that purpose in the cast plate. The forms of the outer side of the chair and the plate, are such as to counteract a side movement of the plate or bolt in an outward direction. The screwbolt and nut weigh about half a pound, and the head of the bolt, which is round, nearly fills a countersink in the bottom of the chair. The intermediate parts of the railway bar, at every 3 or 4 feet, are secured in line upon the string piece, by spikes driven on each side of the rail, with brad heads projecting over the edge of the lower web, in the usual manner of fastening down the H rail. Under the middle of each bar, however, there is introduced a small wrought iron plate, through holes, in which the two opposite spikes, at that place, are driven. This plate was intended to strengthen those spikes, and to aid in securing the proper position of the rail, in the middle of the bar between the chairs, and especially in the curved parts of the line. At the middle of the length of each bar, a notch was cut in one side of the lower web, and in this notch, one of the spikes that passes down through the wrought plate was driven. This is the expedient adopted here to counteract an endwise movement in the bars of the railway. The action is sufficient to bend the spike intended to prevent the movement, and a re-adjustment of the bars, lengthwise, is occasionally rendered necessary. The string pieces and under-sills are of southern pine, and the cross ties are principally of white oak and chesnut, although many are of red cedar, and some are of locust.

MARYLAND. 157

The foregoing is a description of the composition of the track generally; yet many miles of it are laid with timber cut from the adjacent forests. In the parts here alluded to, the iron rails are laid and fastened in the same manner as already described, upon logs of oak and chesnut, hewn upon the upper side, and upon the vertical sides only where the chairs are seated, and also where the logs rest upon heavy cross ties of wood, laid 8 feet apart; the latter as well as the logs upon the cross ties, being imbedded in the road bed.

Upon much of the line the road bed is of sand, and the frost has no injurious effect; in other parts, however, where clay is in contact with the wood of the track, the action of frost heaves the road, but does not derange the line of rails so much as it would, if the bars rested upon cross ties, without the intervention of a string piece.

The average of the actual cost of a mile of this track, comprehending the cost of the lumber, chairs, plates, screw bolts and nuts, spikes, transportation on the common roads, and distribution of the materials, straightening the iron rails and dressing their ends, workmanship in laying the track, turnouts, $50 per ton for 63 tons of iron rails, and superintendence and contingencies, amounted in the aggregate to $7,532.

WILMINGTON AND SUSQUEHANNA RAIL-ROAD, 32 miles in length, commences at the southern terminus of the Philadelphia and Wilmington Rail-road, in the city of Wilmington, and extends to the Susquehanna, opposite Havre de Grace, whence the line is continued to Baltimore, by the Baltimore and Port Deposite Rail-road. The two lines are united by a steam ferry boat, so constructed as to admit the passage of the cars immediately from the rail-road to the dock. Cost $1,200,000.

ANNAPOLIS AND ELKRIDGE RAIL-ROAD, commences on the 18 mile stone from Baltimore, near the Savage factory, on the Baltimore and Washington Rail-road. Leaving the rail-road, the line follows the Patuxent valley, and crossing Chandler's run and the Severn, proceeds to, and terminates at, Annapolis. Whole length of road 19.75. Commenced in July, 1838.

EASTERN SHORE RAIL-ROAD. The route of this work about 170 miles in length, commences at Elkton, and proceeds southward through the counties of Cecil, Kent, Qeeen Ann, Caroline,

Dorchester, Somerset and Worcester, in Maryland, and thence across the state boundary into Accomac county, Virginia, and terminates at King's Creek in Northampton County. A part of the grading of this road has been commenced. Its progress however, being slow, it will be many years before it will be completed.

CANALS.

CHESAPEAKE AND OHIO CANAL, is the work of a joint stock company, chartered by the states of Maryland, Virginia and Pennsylvania, and sanctioned by Congress. The line as originally surveyed, commences at Georgetown on the Potomac, pursues the left bank of that river, passes through the towns of Harper's Ferry, Williamsport, Hancock, and Old town, to Cumberland, where it enters the extensive coal field of Allegany county; thence, leaving the Potomac, it follows the valleys of Wills creek, Cassilmans, Youghiogeny and Monongahela rivers, through Connelsville, and McKeesport to Pittsburg in Pennsylvania. On the finished portion there are 53 locks, 100 by 15 feet, with an average lift of 8 feet; 150 culverts and 7 aqueducts; 6 feeders, formed by as many dams at various points across the Potomac; 60 miles of the canal, from Georgetown upward, is 60 feet wide on the top water line; the remainder is 50 feet wide. The depth (6 feet) is uniform throughout the whole line. On that division of the canal now in progress, extending from Hancock to Cumberland, there will be 22 locks, 40 culverts, 2 dams and 4 aqueducts. About 25 miles above Hancock there is a tunnel, 3118 feet long, 24 feet chord, and 17 feet from the crown of the arch to the water surface, cut through slate rock. There will be another tunnel, through the Allegany mountain, 4.05 miles in length.

The section from Cumberland to Pittsburg has not yet been definitively located. Distance from Georgetown to the Pennsylvania state line, 189 miles. From that point to Pittsburg, $152\frac{1}{4}$ miles; whole length as proposed 341.38 miles; general course N. W.; commenced in 1828; since that time, the work has been steadily prosecuted, and in 1839 that portion of the canal extending from Georgetown to Hancock, a distance of 136 miles, was opened for use. The estimated cost of the work from Georgetown to Cumberland, is $11,591,768 37, of which $1,000,000 were subscribed by the United States, $1,000,000 by the city of Washington, $250,000 by Georgetown, $250,000 by Alexandria, $5,000,000 by the State of Maryland, and $250,000 by Virginia. The entire expenditure from the commencement, to May 31, 1839, has been $8,591,768 37.

WASHINGTON BRANCH of the above extends $1\frac{1}{4}$ miles to the Potomac at Washington.

ALEXANDRIA CANAL, extends from the southern terminus of the Chesapeake and Ohio Canal, at Georgetown, to Alexandria, $7\frac{1}{4}$ miles.

MARYLAND CANAL, OR BALTIMORE JUNCTION with the Chesapeake and Ohio Canal. After a patient and careful investigation on the part of the engineers, of the several proposed routes for this important work, that by way of the village of Brookville, in Montgomery county, was recommended. The summit level of this route is 375 feet above mid-tide, and 16.86 miles in length.

Aggregate length of Rail-roads in Maryland, 262 miles.

Aggregate length of Canals in Maryland, 136 miles.

VIRGINIA.

On the 5th of February, 1816, a Board of Public Works was established in this state, and a fund for Internal Improvements, consisting of revenues derived from stock held by the state in certain canals, turnpikes, banks, &c., was created. This board was authorized to subscribe on behalf of the commonwealth, to such public works, and to such amount, from this fund, as the legislature should from time to time direct; provided three-fifths of the necessary stock of each company shall have been previously taken by other responsible persons.

The board has power to appoint a due proportion of directors; and the state in this and all other respects, is regarded as a stockholder in each case. This arrangement is to continue in operation until the 1st of January, 1866; provided the safety of the commonwealth should not require its suspension. On the 7th of December, 1835, the Internal Improvement Fund amounted to $3,223,484 60.

The most important work now in progress is the James and Kanawha river improvement, and their connection by railroad. It was commenced and prosecuted under the authority of the state until 1835, when it was conveyed to the James river and Kanawha company, in consideration of 10,000 shares of that company's stock transferred to the state. In addition to which the board of public works was authorised to subscribe for $2,000,000 of the stock of the new company.

CANALS.

JAMES RIVER AND KENAWHA CANAL AND RAIL-ROAD. This work is nearly completed, by canal, with some slackwater navigation, between Richmond and Lynchburg, and is under contract between Lynchburg and the mouth of the North River, in Rockbridge county, Virginia. Beyond this point it is yet doubtful what direction it may take. One of the plans being to continue the canal to Covington, and thence construct a rail-road across the Allegany mountains to the Greenbrier, and thence along that stream and the Kanawha, and by way of the valley of the Mud River, and the Guyandotte, to the Ohio, at the mouth of the latter stream. The other to continue the canal only *to Buchanan*, taking the rail-road thence *along the valley of Virginia* to the New River, and thence down this river and the Kanawha to the point of divergence to the valley of the Mud River and the Guyandotte, as above. The latter route seems to be preferable, not only by its greater natural facilities, but because the portion of the rail-road between Buchanan and the New River would, whilst constituting a part of the great western communication of the state, be at the same time so much done of its *south-western* rail-road, a work promising perhaps to the commonwealth even greater advantages than its connection with the Kanawha and Ohio, inasmuch as it would reach a trade not competed for by the lines of improvement through Pennsylvania and Maryland, and which might therefore be secured entirely to Virginia. Whatever direction the James and Kanawha line of improvement may take, it is to be hoped, for the welfare of the state, that its great south-western rail-road will not be long delayed. As soon it shall be completed it may be expected to pour into the canal an immense trade, and to convey on it a very great travel; and until it is effected, the state can scarcely be expected to derive from her investment in the canal, an amount of revenue to justify her expenditure in its construction.

The whole length of the James and Kanawha Canal and Rail-road when completed to the Ohio, will be about 425 miles, that of the proposed South-Western Rail-road from Buchanan to the Tennessee line, about one hundred and sixty miles.

DISMAL SWAMP CANAL, extends from Deep Creek, a tributary of Chesapeake Bay, to Joices Creek, a branch of Pasquotank river of Albemarle Sound; length 23 miles; 46 feet wide, $6\frac{1}{2}$ deep, at intervals of a quarter of a mile, the canal is widened to 60, for turn-out stations; 6 locks 100 by 20 feet; summit level $16\frac{1}{2}$ feet above the Atlantic at mid-tide. Two lateral canals, one from Lake Drummond, 5 miles in length, which, in addition to its uses for the purposes of navigation, serves as a feeder to the main trunk; and the other, 6 miles long, opens a communication between the principal canal, and the head waters of North West river.

The navigation of the Roanoke, Rivanna and Slate rivers has been partially improved by means of dams and locks. Companies have been incorporated for the purpose of improving the navigation of the Shenandoah, Catawba, Nottoway, Upper Banister, Tuckahoe, and South Anna; Coal, Smith's, Cowpasture, Goose, North Anna, and Pamunkey rivers.

RAIL-ROADS.

The principal works of this description in the state, are the Richmond and Fredericksburg, Richmond and Petersburg, and Petersburg and Roanoke Rail-roads, immediately on the line of northern and southern travel through the state, and constituting portions of the great northern and southern line of communication so rapidly progressing to completion between New York and New Orleans. When Virginia shall have completed the rail-road between Fredericksburg and the Potomac, and corrected the roads at Richmond and Petersburg, she will have done every thing that can reasonably be expected of her, in facilitating the traveller on this important line of thoroughfare. A rail-road between Washington and the termination of the Richmond and Fredericksburg Rail-road on the Potomac, would still be required to prevent the interruption which now exists, when the Potomac is frozen; but such a work seems to be rather an object of national than of state concern, or at all events one in which the national government, so deeply interested in its execution, may reasonably be expected to co-operate with Virginia towards its construction. The following details in relation to the works above named, will probably be of interest.

RICHMOND, FREDERICKSBURG, AND POTOMAC RAIL-RAOD. This work lies in Henrico, Caroline, and Spotsylvania counties, between Richmond and Fredericksburg. It is 61 miles long, and when extended to the Potomac, at the mouth of Aquia Creek, (at which point it is proposed to terminate it) will be 75 miles long. It has on it several fine bridges at the crossings of the North and South Anna rivers, Little river and Chickahominy, and will pass the Rappahannoc at Fredericksburg by a bridge forty feet high and six hundred feet long, and Potomac Creek, by a bridge three hundred feet in length, and seventy feet high. The grades and curves of the road are in general favourable, the sharpest of the latter being about 2000 feet radius, and the maximum graduation, with a single exception near Fredericksburg, (where a grade of forty-five feet was necessary) being about thirty feet per mile. The execution of the work is, with the exception of the superstructure, highly substantial. The superstructure is of the ordinary wooden rail, plated with iron, in general use in the south, and recommended on this and most other southern roads, by the cheapness of timber and motives of economy. When the travel on this road, which is increasing very rapidly, shall have become larger, the company will probably find it to their advantage to lay down heavy iron rails the whole length of their road. With this addition, it would be one of the best, as it promises to be one of the most productive in the country.

The whole cost of the Richmond and Fredericksburg Railroad, including a branch four miles long to the Deep Run coal mines, and including also a full supply of locomotives, cars, &c., has been, between Richmond and Fredericksburg, $1,100,000. An additional expenditure of $270,000, is estimated as requisite to complete the rail-road to the Potomac.

RICHMOND AND PETERSBURG RAIL-ROAD. This rail-road, 23 miles long, connects the towns of Richmond and Petersburg. Its curves and grades are both highly favourable, a large proportion of the line being straight, and the maximum graduation being thirty feet per mile. The rail-road bridge across the James river, at Richmond, on this road, is one of the most striking works in America, being about 3000 feet long, between 60 and 70 feet above the river, and having spans of 150 feet and upwards. The cost of this viaduct was but $125,000, and the whole cost of the rail-road, including depots, locomotive engines, cars, &c. about $750,000.

A branch of ten miles in length has been proposed from this rail-road to Bermuda Hundred, which is said to be of extremely easy execution, and which would probably add so much to the profits of the rail-road, and to the commercial facilities of

Richmond, that it is a matter of surprise it has not yet been executed. By means of it, produce would be delivered on board European vessels in two hours after leaving the warehouse, instead of being, as at present, delayed a day or two on a circuitous navigation, in lighters; and merchants would be able to communicate quickly and constantly with their ships. This slight improvement, and the removal of Harrison bar, would do much to make Richmond an important sea-port.

PETERSBURG AND ROANOKE RAIL-ROAD. This work, 59 miles long, extends from Petersburg, through the counties of Chesterfield, Dinwiddie, Sussex and Greensville, to the Roanoke river, opposite Weldon. Near its termination, it connects with the Wilmington and Raleigh Rail-road in North Carolina; and it is connected by means of the Greensville Rail-road with the Raleigh and Gaston Rail-road at Gaston. This rail-road has on it no curve, after leaving Petersburg, of less radius than half a mile, and no grade, except at that point, exceeding 30 feet per mile. Like the Richmond and Fredericksburg, and Richmond and Petersburg Rail-roads, its roadway formation, bridges, &c. are executed in a highly substantial and permanent manner, but it has only a plate-rail superstructure, which with the increasing business of the road, it will probably be advisable to substitute by a substantial and heavy iron rail. The whole cost of this rail-road, including depots, locomotive engines, cars, &c. has been only $800,000. By this rail-road a large portion of the produce of the Roanoke valley is conveyed to Petersburg.

Besides the above rail-roads, on the great northern and southern line of travel through Virginia, there are the following:—

THE GREENSVILLE RAIL-ROAD. This rail-road, as above observed, connects the Petersburg and Roanoke with the Raleigh and Gaston Rail-road. It leaves the former rail-road a few miles south of Hicksford in Greensville county, Va., and terminates at Gaston in North Carolina, where the Raleigh and Gaston Rail-road commences. It is 18 miles long, and cost about $250,000. Its construction is similar to that of the Petersburg and Roanoke Rail-road.

CITY POINT RAIL-ROAD. This rail-road extends from Petersburg, along or near the Appomatox, to City Point, below the junction of that stream with the James river. It is twelve miles long, and has cost about $200,000. By means of it an important facility is afforded to the shipping interests of Petersburg, which have hitherto been dependent on lighters on the Appomatox.

CHESTERFIELD RAIL-ROAD. This rail-road connects the bituminous coal basin of Chesterfield county with the tide water of James river at Manchester, opposite Richmond. Length 13.50 miles, with branches from the main line to the principal coal pits. The line descends into the valley of Falling Creek by a self-acting plane, and is conducted over that of Sally's Run by an embankment 800 feet long and 40 feet high. Cost but $8,000 per mile. Radius of the curves, 1442 feet. Single track, with the usual turn-outs, &c.; commenced January, 1830; completed July, 1831.

LOUISA RAIL-ROAD. This rail-road leaves the Richmond and Fredericksburg Rail-road, about 24 miles from Richmond and 37 from Fredericksburg, and extends thence westwardly to Gordonsville in Orange county, passing by Louisa Court-house and Newark. Whole distance from the point of junction with the Richmond and Fredericksburg Rail-road to Gordonsville, 49 miles. Cost about $400,000.

The Louisa Rail-road furnishes an outlet to the productions of a fertile tract of country in Virginia at the foot of the South-West Mountain, and forms the most convenient and agreeable route to travellers to the Virginia Springs and the western part of Virginia. It is proposed to extend a branch of the road from Newark, about 12 miles east of Gordonsville, to Charlottesville; and an extension of the road to Harrisonburg, in Rockingham county, is spoken of.

PORTSMOUTH AND ROANOKE RAIL-ROAD. This rail-road commences at Portsmouth opposite Norfolk, and proceeds in a W. S. W. direction, through Norfolk, Nansemond and Southampton counties, crosses the North Carolina boundary near Meherrin river, and joins the Roanoke near Weldon. Length 80 miles. Cost about $1,000,000.

WINCHESTER AND POTOMAC RAIL-ROAD, extends south-west from Harper's Ferry on the Potomac, in Jefferson county, Va., where it connects with the Baltimore and Ohio Rail-road, to Winchester in Frederick county. Length 32 miles. Cost $500,000. From Winchester a Macadamized road is now under construction to Staunton; but it is probable that this will ere long be superseded by a rail-road, which would be so well justified along the fertile valley of Virginia, remarkable for both its mineral and agricultural wealth. A company for the construction of this rail-road, was incorporated by the legislature of Virginia a few years since, and a survey made, and very favourable location procured for the work; but the capital stock was not made up within the time limited in the charter for the organization of the company.

The above comprise all the rail-roads yet executed in Virginia. Other highly important works of this description have been projected in that state, which will probably be undertaken as soon as the present monetary difficulties pass off. One of the most essential seems to be a rail-road between Richmond and Lynchburg, along the ridge dividing the waters of the James and Appomatox rivers, to be extended to connect with the proposed south-western rail-road. Such a rail-road would secure to the state a large travel and return trade to the south-west, in light and valuable packages, which, without it, will probably not pass through the state. A second rail-road, which would add largely to the wealth of the country through which it would pass, and to the commerce of Richmond, has been surveyed, diverging from the former in the neighbourhood of Farmville, and passing in a south-westerly direction to the valleys of the Staunton and Dan rivers, a short distance above the junction of these streams. With her great western and south-western improvements carried out, and these two lines of rail-road, Virginia might again take rank among the foremost states of the confederacy.

Rail-roads have been proposed from Weldon, in North Carolina, up the valley of the Roanoke, through Clarksville, to Danville, in Pittsylvania county. From Danville, via Martinsville, to the Kanawha, in Wythe county, where it will intersect the line from Lynchburg to the Tennessee boundary.

From Staunton to Potomac river. From Staunton to Scottsville. Along the valley of the Rivanna. From Richmond to Danville, via Farmville, Banister, &c. From Fredericksburg to a point on the Staunton and James River Rail-road, via Orange Court House and Charlottesville. From Petersburg to Farmville. From Suffolk to the Portsmouth and Roanoke

VIRGINIA. 167

Rail-road. From Taylorsville to Louisa Court House, and Orange Court House. From Lynchburg to the Tennessee line, via Abingdon. From Cherrystone to the Maryland line. From Richmond to Yorktown. From Warrington to the Falmouth and Alexandria Rail-road. From Smithfield to the Winchester and Potomac Rail-road. From Kanawha Salines to Coal river, and some others.

Although companies have been chartered by the legislature for the construction of these roads, many of them it is probable will never be executed. The aggregate amount of capital of these companies, exceeds twenty-five millions of dollars.

Aggregate length of canals in Virginia 196.25 miles.
"   "   rail-roads   "   361.50 miles.

NORTH CAROLINA.

RAIL-ROADS.

WILMINGTON AND RALEIGH RAIL-ROAD. This road extends from Wilmington to Weldon on the Roanoke river, and connects with the Portsmouth and Roanoke Rail-road, and the Petersburg road. It is 161 miles long, is completed, and in active operation.

According to the original charter of 1833, this company was required to construct a rail-road from Wilmington to Raleigh, but by an amendment to the act of incorporation, passed in 1835, when its capital stock was increased from $800,000 to $1,500,000, the company was authorized to change its direction from the former place to some point on the Roanoke. Hence the anomalous appellation of " Wilmington and *Raleigh* Rail-road." By the new location the road does not approach within fifty miles of Raleigh. It commences at the town of Wilmington in New Hanover county, and passes by a course nearly north, through Duplin, Lenoir, Green, Edgecombe and Halifax counties, and terminates at Weldon. The road was commenced in October 1836, and completed 7th March, 1840. $21\frac{1}{2}$ miles consist of curves, and $138\frac{1}{2}$ of straight lines; one of these straight lines is 47 miles in length. The minimum radius of curvature is 3730 feet, most of the radii are 12,200 and 30,000 feet. The radius of one curve is 67.240 feet, which is deemed for all practical purposes, equivalent to a straight line. Maximum inclination, 30 feet per mile, but nearly all the gradients are level.

RALEIGH AND GASTON RAIL-ROAD, extends from Gaston on the Roanoke, where it unites with the Petersburg, Greensville and Roanoke Rail-roads, and terminates at Raleigh, the capital of N. Carolina, a distance of 85 miles, passing through the counties of Halifax, Warner, Granville, Franklin and Wake. It is completed to Tar river, about 35 miles from Gaston; the balance of the road is rapidly progressing towards completion, and the whole is expected to be opened for travel in the course of the present year (1840.) From Raleigh it is proposed to continue it to Columbia in South Carolina, for which a charter has been obtained, thus forming an uninterrupted communication by rail-roads from Fredericksburg in Virginia, to the latter place.

A RAIL-ROAD from Raleigh to Columbia in South Carolina, from Fayetteville to the Narrows of the Yadkin, with a Branch thence to the Louisville, Cincinnati and Charleston Rail-road, and another to Wilkesboro are proposed.

For an account of the Rail-roads extending from the Roanoke northward, see Virginia.

CANALS.

DISMAL SWAMP CANAL. (See Virginia.)

LAKE DRUMMOND CANAL. A navigable feeder of the preceding; it extends from Lake Drummond to the summit level of the Dismal Swamp Canal; length 5 miles; 16 feet wide; 4 and a half deep, with a guard gate near the lake.

NORTH-WEST CANAL, connects North-West river with the Dismal Swamp Canal; length 6 miles; 24 feet wide, 4 feet deep.

WELDON CANAL, forms the commencement of the Roanoke

navigation. It extends around the falls of Roanoke, above the towns of Weldon and Blakely; length 12 miles; lockage 100 feet.

CLUBFOOT AND HARLOW CANAL, extends from the head waters of Clubfoot, to those of Harlow creek, near Beaufort; length $1\frac{1}{2}$ miles.

The navigation of the Roanoke from the Weldon canal to the town of Salem in Virginia, a distance of 232 miles; the Cape Fear, the Yadkin, the Tar, New and Catawba rivers, has been greatly improved by joint stock companies.

Aggregate length of Rail-roads in North Carolina, 250.00 miles.

Aggregate length of Canals in North Carolina, 13.50 miles.

SOUTH CAROLINA.

RAIL-ROADS.

SOUTH CAROLINA RAIL-ROAD, commences at Charleston, pursues a north-west course, and crosses the head waters of Ashly river, 28 miles from Charleston; 7 miles farther it crosses Four-holes swamp. At a distance of 65 miles from Charleston the Edisto is passed : thence by a direct course, 58 miles, it enters the valley of Big Horse Creek, which it pursues for a few miles, then runs westward, and terminates in the town of Hamburg, opposite Augusta. Entire length, 135.75 miles. Several towns and villages have been erected along the line of this road ; among them are Beesville, Summerville, Branchville, Midway, Blacksville, Aken, &c.

The plan of the road is unusually straight, and the curves have large radii. The profile is gently undulating, frequently nearly level, and the maximum ascent does not in any case exceed 30 feet to the mile. The summit of the dividing ridge, between the Savannah and Edisto, is elevated 513 feet above the tide; and one inclined plane, (the only one on the line,) provided with a stationary steam-engine, is resorted to at this spot, which is 114 miles from Charleston. The superstructure is composed of flat iron bars, attached to wooden string-pieces, 6 × 10 inches, supported generally on piles; the latter are secured by ties, and are sometimes of a great length. They have been driven to a considerable depth in some of the marshes which the road crosses, and in other parts of the work they form a substitute for embankments, which latter have not been resorted to, except in a few very limited situations. The railway resembles a continuous and prolonged bridge. Stones are not employed on this line, for two reasons—first, the country is completely destitute of this material—secondly, it is not necessary in the mild climate of the south. (The natural earth, when dry and not exposed to frost, forms a firmer foundation or support, for piles, sleepers, stone blocks or stone sills, than the irregular projections of broken stone. The only use of the latter material is for the purpose of draining off the water, by permitting it to sink among the crevices of the stones until it reaches the bottom of the pit or trench, which is below the usual depth to which frost can penetrate; from this trench the water is conducted by drains, or is absorbed by the earth.) The exposed parts of the wood work have been protected by a coating of heated tar and oil. Where the foundation is a uniformly hard clay, transverse sleepers are firmly bedded in and on the clay, for the support of the rail timber; where the excavated surface is of a less firm character, foundation timbers, running parallel to the road, are bedded in the earth, on which the transverse caps, which support the rail timber, are secured; and where the foundation is too loose or yielding to allow sleepers, or the line of graduation is above the surface, piles, driven into the ground, are made use of to support the caps. Cost $1,750,000; commenced in 1830; completed in 1834; since sold to the Louisville, Cincinnati and Charleston R. R. Co. for $2,400,000. The receipts on the Charleston and Hamburg Rail-road, for the month of November, 1839, amount to $65,000, nearly 50 per cent. more than was taken in any previous month.

BRANCHVILLE AND COLUMBIA RAIL-ROAD, extends in a north-west direction from Branchville, on the South Carolina

SOUTH CAROLINA. 171

Rail-road, in Colleton District, 62 miles from Charleston, through Orangeburg District, to Columbia, in Richland, a distance of 66 miles. This road forms a part of the contemplated Louisville, Cincinnati and Charleston Rail-road. Maximum grade 25 feet per mile; minimum radius of curvature 2,800 feet; cost $1,500,000.

LOUISVILLE, CINCINNATI AND CHARLESTON RAIL-ROAD. A company, with banking privileges, was incorporated in 1836, by the legislatures of South Carolina, North Carolina, Tennessee and Kentucky, for the purpose of constructing a rail-road through their respective states. A portion of the South Carolina Rail-road, extending from Charleston to Branchville, a distance of 62 miles, and the branch of that road from the latter to Columbia, 66 miles in length, have been adopted as a part of the great north-western line. From Columbia, it is proposed to continue the road along the valleys of Broad River, in South Carolina, and of the French Broad River, of North Carolina and Tennessee, to Knoxville, in the latter state; thence through Kentucky to Newport, on the Ohio River, opposite Cincinnati.

As a preparatory step in this great enterprise, the purchase of the South Carolina Rail-road was effected for $2,400,000, paid principally in the stock of the new company, whose capital is $8,000,000. Entire length from Columbia to Cincinnati, about 590 miles; and from Charleston, 718 miles. With regard to the execution of this, nothing is yet done beyond Columbia.

Another route to the same points, has been urged upon the attention of the company. It embraces the following Railroads, some of which are now in operation, and others in progress:—Georgia Rail-road, from Augusta to De Kalb county, in Georgia; Western and Atlantic Rail-road, from De Kalb to Chattanoogo, on the Tennessee and the Highwassee Railroad, from Chattanooga to Knoxville, the latter place being common to both routes. Distance from Charleston to Cincinnati by this route, 741 miles.

CANALS.

SANTEE CANAL, connects the harbour of Charleston with the Santee. It commences on the west branch of Cooper river, and passing along Biggin Swamp, intersects Santee river at Black Oak Island. Length, 22 miles; course, N. N. W.; 32 feet wide at top, 20 at bottom; 4 feet deep; rise and fall, 103; 13 locks, each 60 by 10 feet; completed in 1802; cost, $700,000. By means of this canal, and the Santee and Congaree rivers, which have been improved, a navigable communication is afforded from Charleston to Columbia.

WINYAW CANAL, extends from Winyaw Bay to Kinlock Creek, a branch of Santee river. Length, 7.40 miles; course, S. W.

THE NAVIGATION OF THE CATAWBA, has been improved by the construction of several small canals:—1, extends from Patton Island to Davy's Ferry, 2 miles; 2 from $1\frac{1}{4}$ miles below M'Donald's Ferry to Fishing Creek, $2\frac{1}{4}$ miles; 3, from Mountain Island to Rocky Creek, $1\frac{3}{4}$ miles; 4, from Rocky Creek to the Catawba, 900 yards; 5, from Jones's Mill to Elliot's, 4 miles, (this is styled the Wateree canal.)

SALUDA CANAL, extends from the head of Saluda Shoals to Granby Ferry, on the Congaree, passing through the town of Columbia; 6.20 miles in length; descent 36 feet.

DREHR's CANAL, is designed to overcome a fall of 120 feet in Saluda river. Length, $1\frac{1}{2}$ miles.

LORICK's CANAL, on Broad river, $1\frac{1}{2}$ miles above Columbia; 1 mile long.

LOCKHART's CANAL, in Union District, around Lockhart's Shoals in Broad river, $2\frac{3}{4}$ miles long.

Aggregate length of canals in South Carolina, 52.45 miles.
"   rail-roads   "   201.75   "

GEORGIA.

RAIL-ROADS.

Georgia Rail-road, commences at Augusta, the head of navigation on the Savannah river, and proceeds nearly due west to the vicinity of Warrenton; thence, curving towards the north-west, and passing through Crawfordsville, Greensboro, Madison and Covington, it terminates at a point a little to the south-west of Decatur, in De Kalb county. At this point the Western and Atlantic Rail-road continues the line to the Tennessee, intersecting, in its course, the Highwassee Rail-road which extends to Knoxville. Length of the Georgia Rail-road from Augusta to its termination in De Kalb county, 165 miles. The first 57 miles of this road are laid with wooden superstructure, and a heavy plate rail $2\frac{4}{10}$ inches wide by $\frac{8}{10}$ thick; the remainder is built with a T rail, weighing 46 lbs. per yard. The inclination of the grades does not exceed 36 feet per mile. Cost, including branch to Athens, $3,300,000.

Athens Branch of the Georgia Rail-road, leaves the latter at a point nearly equidistant between Crawfordsville and Greensboro, and extends a north-west course to Athens, in Clark county. Length 33 miles.

Western and Atlantic Rail-Road, commencing at the point of termination of the Georgia Rail-road, near Decatur, the road crosses the Chattahooche, and ascends to Marietta, in Cobb county. It there crosses the Kermesaw Summit on the north side of the mountain of that name, and descends towards the Etowah, passing through the village of Allatoona. Having crossed the Etowah, the road passes through Two Run Gap, enters the valley of Conasseen's Creek, and ascends in this valley to the Oothocaloga Summit. Thence, by a pretty direct route, to the Oostanauley, which the road crosses, and proceeds towards its termination on the Tennessee river at the confluence with the Chickamanga. It branches before it reaches the Tennessee line and unites with the Highwassee Rail-road. Length 130 miles. Cost $2,129,920. Single track, but graded for two. Maximum grade 30 feet per mile. Minimum radius of curvature 1200 feet.

CENTRAL RAIL-ROAD, commences at Savannah, proceeds in a direct course towards the Ogeechee river whose valley is followed; and passing within four miles of Sandersville, terminates at Macon. Length 193 miles; maximum grade 30 feet per mile; minimum radius of curvature 2000 feet. Estimated cost $2,300,000, of which $1,187,032 55 had been expended on the 1st of November, 1839. Single track with turn-outs, &c. This road is rapidly verging towards completion, 80 miles being finished, 48 nearly so, and 28.50, extending to the Oconee, under contract. The construction of the railway is similar to that generally adopted in the south.

The advantages of a continuous bearing, by means of the broad string piece laid flat, are apparent on this road. In colder climates, where it is necessary, and even unavoidable that the foundation should be laid so low as to be out of the reach of frost, such a bearing might not be admissible; but they have nothing to guard against on this score; it is, therefore evident, that the nearer the foundation is laid to the surface of the grade, the more accessible it is for the purpose of repair, renewal or adjustment.

The objection commonly urged against this plan of superstructure, arises from an apprehension, that the ribbon which immediately supports the plate rail, will give way and be crushed by the weight of the engine. Burden and passenger trains have been running over this road daily for more than 18 months, and for some time past, from two to three trains per day, and with the exception of the renewal of the ribbon for a few miles on the lower end of the road for the purpose of substituting a different kind of connecting plate, there has not been one-tenth of a mile renewed for the whole distance of 80 miles. The sides of the embankments are becoming covered with vegetation, and will in a year or two be entirely protected from the effect of rains.

The allignment of the road for the distance located, comprises 61 curved, and 62 straight lines.

GEORGIA. 175

The curves are all arcs of circles and may be classed as follows:—

Length of Radius.	Number of Curves.	Aggregate distance.
2,000 feet.	14	24,359 feet.
2,500 "	3	6,608 "
3,000 "	2	4,086 "
3,500 "	3	7,435 "
4,000 "	7	15,369 "
4,500 "	2	4,980 "
5,000 "	12	40,472 "
8,000 "	4	12,984 "
10,000 "	6	23,405 "
15,000 "	5	21,916 "
20,000 "	1	8,374 "
30,000 "	1	4,620 "
150,000 "	1	26,500 "
Total	61	201,109 feet.

Total length of curved line,   38 miles and 469 feet.
  "   "   straight line,   110 "   " 2,591 "

Distance located   148 miles   3,060 feet.

The last mentioned curve of 150,000 feet radius, and about five miles in length is, so far as any effect of resistance is considered, fully equivalent to a straight line, at any velocity. We may therefore, with propriety, state the proportion of straight line at two-thirds of the whole distance.

The gradients may be classed as follows:

	miles.	feet.
Level	20	2200
Inclination of 5 feet per mile and under,	43	1560
over 5 and under 10	22	3440
over 10 and under 15	14	5180
over 15 and under 20	12	940
over 20 and under 25	8	2360
over 25 and under 30	26	3020
Total	145	2860

The arrangement of the curves and slope grades, is such as to avoid, excepting in a few instances, the occurrence of a sharp curve on a heavy grade.

Water stations are established 10 miles apart, or as near this distance as the circumstances will permit. At each station is a "turn-out" about 800 feet in length, to allow two trains to pass each other. It is presumed that it will at a future day become necessary to place "turn-outs" intermediately between the present ones.

In most instances store-houses will also be erected at the stations for the accommodation of the local business—and dwellings for the persons entrusted with the supervision of the road.

The business of this road for the three months, ending 31st October, 1839, was as follows:—

	Passengers.		Freight.		Aggregate.	
	No.	Amount.	Dolls.	Cts.	Dolls.	Cts.
Aug.	747	$1,464 25	2,108	77	3,573	02
Sept.	688	1,565 10	6,278	49	7,843	59
Oct.	875	2,215 50	11,844	99	14,060	49
	2310	$5,244 85	$20,232	25	$25,477	10

AUGUSTA BRANCH of the Central Rail-road. A charter was granted at the last session of the legislature, for a branch railroad to connect the above road with the city of Augusta; and in compliance with a request from a committee of the citizens of Burke county, a survey was made for the purpose of ascertaining the cost, &c. of that portion of the route between the Central Rail-road and Waynesboro. A report with estimates and maps in detail, showing the result of this survey, was communicated to the above-named committee. As that report has not been published, the following synopsis may be made:

The route surveyed diverges from the line of the Central Rail-road about three-fourths of a mile below the point where this road crosses Big Buckhead creek, and pursues the general direction of the valley of this creek for about 13 miles, to Rosemary creek—here bending to the right, it assumes the dividing

GEORGIA. 177

ridge between the waters of Buckhead and Briar creeks, and follows this ridge over a moderately undulating country to Waynesboro.

The distance is $22\frac{1}{2}$ miles—which, added to the distance from the point of junction to the city of Savannah, 79 miles—and the distance from Waynesboro to Augusta, $32\frac{1}{2}$ miles—makes a total distance of 134 miles from Savannah to Augusta by rail-road, being only 12 miles longer than the direct stage route.

There will be no inclination of grade exceeding 30 feet per mile, and no curvature on a radius of less than 2000 feet.

The cost of the road from the Central Rail-road to Waynesboro is estimated at $182,800, exclusive of locomotive engines, cars, &c.; and contemplating a superstructure similar to that of the Central Rail-road, with a plate rail supported by longitudinal string-pieces.

The citizens of Savannah, by an unanimous vote in town meeting, requested the corporate authorities to subscribe $100,000 to the capital stock of this road, and should the city of Augusta take a like sum, there is every reason to expect that the large resources of the county of Burke, and the public spirit of its citizens, with those of the two cities, will supply the remainder of the required funds, and that we shall soon see this branch in progress.

That it would be of great advantage to the cities of Augusta and Savannah and the intervening country, and add greatly to the business of the two rail-roads already in progress, no one will doubt; and that the estimated cost bears a small proportion to the great advantages and revenue that might be expected, will also be readily admitted.

MONROE RAIL-ROAD, is a prolongation of the Central Rail-road, extending from Macon to Forsyth, 25 miles in length; and thence to De Kalb, where it unites with the Western and Atlantic Rail-road. Its structure is similar to that of the Georgia Rail-road. Maximum grade 3696 feet per mile. Minimum radius of curvature 1910 feet. Single track, with turn-outs, &c. Cost $20,000, per mile.

MACON AND TALBOTTON RAIL-ROAD. This is an extension of the Central Rail-road to Talbotton, whence it is proposed to extend a branch to West Point, where it will unite with the

GEORGIA.

Montgomery and West Point Rail-road of Alabama, and another to Columbus on the Chattahoochee; 70 miles in length.

COLUMBUS AND CHATTAHOOCHEE RAIL-ROAD, leading from Columbus, via West Point, to the south-eastern terminus of the Western and Atlantic Rail-road in De Kalb county.

BRUNSWICK AND FLORIDA RAIL-ROAD, to extend from Brunswick on the Atlantic coast, to a point on the Appalachicola, and thence to Choctawhatchie Bay.

CHATTAHOOCHEE RAIL-ROAD, from Macon to West Point, via Columbus.

CANALS.

BRUNSWICK CANAL, 3¼ excavation finished, from tide-water on the Alatamaha to Brunswick, 12 miles; cost $500,000; thorough cut with tide locks.

SAVANNAH, OGEECHEE AND ALATAMAHA CANAL, is undergoing repairs. The lock and canal near Savannah is to be made 100 feet wide and 12 feet deep; 16 miles long; commenced in 1825; completed in 1829; cost $165,000.

Aggregate length of rail-roads in Georgia, 616.00 miles.
"   "   canals   "   28.00 "

FLORIDA.

RAIL-ROADS.

LAKE WIMICO AND ST. JOSEPH'S CANAL AND RAIL-ROAD, 12 miles in length. Incorporated in 1835; work completed 1836.

EAST FLORIDA RAIL-ROAD, proposed, and surveys made. It extends from Jacksonville, on the river St. John's, to St. Mark's. Estimated cost $1,233,000.

RAIL-ROADS, from Tallahassee to St. Mark's, 20 miles; from Columbus, Georgia, to Pensacola Bay, 120 miles; from Jacksonville to Tallahassee, 150 miles; from Pensacola Bay to Mobile Bay, 40 miles; from St. Joseph's to Tallahassee, 70 miles; from St. Augustine to Picolata, 18 miles; are proposed.

CANALS.

CANALS from Matanzas River to Halifax River, 15 miles; from St. Andrew's Bay to Chipola, are proposed.

ALABAMA.

RAIL ROADS.

ALABAMA, FLORIDA AND GEORGIA RAIL-ROAD, commencing at the intersection of Broadway and Hancock streets in Pensacola; the line proceeds northwardly to the western bank of the Escambia. From thence the Escambia valley is passed at nearly right angles, crossing the main Escambia and some of its tributaries, to the main land on the eastern shore. Ascending and following the dividing ridge between Escambia and the streams of Blackwater bay, the line reaches the eastern declivity of Conecuh valley. Thence, descending the eastern slope, and passing a few miles up the valley, it crosses to the western side of the Conecuh, and proceeds to the Sepulga. Thence, deflecting to the north, it follows the valleys of the Sepulga, Pigeon creek and Three Run to the summit, 417 feet above tide, between Mobile and Pensacola bays. Thence descending by the valley of Pinchoma, and crossing the head waters of Pintlala and Catoma creeks, the line terminates at Montgomery, in Montgomery county, Alabama. The road ascends from the valley of Escambia, by a grade of 35 feet per mile, for 4.15 miles, and thence along the ridge 22.51 miles with an undulating graduation, varying from 15 to 30 feet per mile, though it seldom attains the latter inclination. The descent to the Conecuh valley for 4.77 miles is 37 feet per mile, thence by a diminished grade, it reaches the crossing of the Conecuh. From this point the surface is highly favourable, seldom requiring an inclination beyond 20 feet per mile. From the summit the maximum grade is 29 feet per mile for the entire length of Pinchoma valley. The radius of the curves throughout the whole line is 6180 feet. The aggregate of straight line is

130.46 miles; that of curves, 26 miles. Entire length from Pensacola to Montgomery, 156.46 miles.

The road is constructed for a single track, with turn-outs and side lines. The levels are attained by the ordinary modes of excavation and embankment, except over deep ravines and marshes, where trestle bridging or piling will be resorted to. The Escambia marshes are crossed by piles, driven 6 feet apart from centre to centre, longitudinally and transversely with the line of the road; the piles are cut off 4 feet above the level of the marsh, and their ends connected by cross-caps 9 feet long and 10 inches square. Longitudinal string pieces, 10 inches square, are notched into the cross-caps and bolted through. Iron rails $2\frac{1}{2}$ inches wide, with sufficient depth to permit the flanges of the car wheels to pass clear of the wood, are secured along the centre of the string-pieces. Sills from 30 to 40 feet long, flattened on two sides to 8 inches of heart pine, are laid in trenches longitudinally with the road, 5 feet apart from centre to centre. The top surface of the sills conform with the grade, and their ends meet on a splicing sill to prevent unequal settling. Cross-ties, 8 feet long, and 8 inches square, are laid upon the longitudinal sills, 4 feet apart from centre to centre. The cross-ties are notched 3 inches deep, for receiving longitudinal string-pieces or rails, 6 by 8 inches square, lined on top, and framed where they enter the notches, into which they are secured by wooden wedges. The notches in the sills are so arranged that when the string-pieces are keyed in, their interior faces are 4 feet $6\frac{1}{2}$ inches apart. The interior top angle of each-string piece is hewn off 1 inch, leaving the track 4 feet $8\frac{1}{2}$ inches wide. Iron rails $2\frac{1}{2}$ inches wide, and $\frac{3}{4}$ of an inch thick, with mitred ends and countersunk holes, are laid flush with the inner faces of the string-pieces, upon which they are secured by spikes $\frac{7}{1}$ of an inch square, 7 inches long at the end holes, and $5\frac{1}{2}$ inches for the intermediate points, to prevent the rails from sinking into the wood; they are supported at the points of junction on iron splicing plates, 6 inches long, and $\frac{1}{4}$ of an inch thick. The wooden rails are adzed down outside the iron, to pass off the rain water, and the track ballasted with sand or gravel to give additional stability to the structure. The average cost per mile of rail-road superstructure, as above specified, may be estimated at $5,500. Total cost of road, buildings, apparatus, &c. $2,500,000.

SELMA AND CAHAWBA RAIL-ROAD. This road, a branch of the Pensacola and Montgomery Rail-road, leaves the main line, and proceeds in a western direction, and at a point 10 miles from Selma and 9 from Cahawba, branches off towards those places respectively.

MONTGOMERY AND WEST POINT RAIL-ROAD. This rail-road extends from the northern terminus of the Pensacola and Montgomery Rail-road, in an east-north-east direction, to West Point, at the head of the rapids of the Chattahoochee river, and about 30 miles above the town of Columbus. It is 87 miles in length, and forms the connecting section between the Georgia and Alabama systems of rail-road. It enjoys an exclusive monopoly of the trade, by rail-roads and locomotives, in the region between the waters of Alabama and Chattahoochee rivers, for fifty years.

WETUMPKA RAIL-ROAD, is designed to connect the Tennessee with the Alabama, at Wetumpka, at the head of steam-boat navigation, in that river. The first object of the company is to construct a rail-road from Wetumpka, along the valley of the Coosa, to Fort Williams, near the head of the great falls, distant about 56 miles. To avoid the obstructions in the river, the road will be extended so as to unite with the Selma and Tennessee Rail-road, and the Georgia Rail-road.

MOBILE AND CEDAR POINT RAIL-ROAD, 28 miles in length.

SELMA AND TENNESSEE RAIL-ROAD, is designed as the great central line of communication between North and South Alabama. The line as surveyed, commences at Selma, on the Alabama, and passes by a line a little east of north, to Gunter's landing, at the most southern bend of the Tennessee river. The length about 170 miles. A branch along the valley of the Coosa, to intersect the Western and Atlantic Rail-road of Georgia, near Echota, is proposed.

CAHAWBA AND MARION RAIL-ROAD, extends up the right declivity of Cahawba valley. Length about 35 miles.

TUSCUMBIA, COURTLAND AND DECATUR RAIL-ROAD, extends from Tuscumbia to Decatur, both on the Tennessee river, 44 miles; maximum inclination, 28 feet per mile. It is contemplated to extend this road into Georgia, and unite it with the rail-roads of that state. Should this be done, and the Memphis

ALABAMA. 183

and La Grange Rail-road of Tennessee be carried to Tuscumbia, as proposed, a complete line of rail-roads will be established from the western extremity of Tennessee to Charleston in South Carolina, a distance of 709 miles.

CANALS.

MUSCLE SHOALS CANAL, extends along, and is intended to overcome the obstruction in the Tennessee, called the Muscle shoals. The work is not yet complete, as an important section at the head of the shoals remains to be executed. When finished, this canal will open an uninterrupted steam-boat channel for several hundred miles into the rich agricultural and mineral districts of East Tennessee. The locks on this canal are 32 feet wide, and 120 in length. Length of the section now in use, 35.75 miles; 60 feet wide at top, 42 at bottom, and 6 feet deep; 16 lift and 2 guard locks, overcoming an ascent of 96 feet; cost $571,835. Estimated cost of all the improvements from Brown's Ferry at the head of the shoals to Florence, $1,361,057.

A Canal around Colbert's shoals is proposed.

HUNTSVILLE CANAL, from Triana on the Tennessee, to the town of Huntsville; 16 miles in length.

Aggregate length of Rail-roads in Alabama, 307.46 miles.
Aggregate length of Canals in Alabama, 51.75 miles.

MISSISSIPPI.

RAIL-ROADS.

WEST FELICIANA RAIL-ROAD, extends from St. Francisville, on the left bank of Mississippi, to Woodville, in Wilkinson county, Miss. Length, 27.75 miles; 7.75 miles being in Mississippi, and 20 in Louisiana; capital, $500,000.

VICKSBURG AND CLINTON RAIL-ROAD, commences at Vicksburg, in Warren county, and proceeds to Clinton, in Hinds county, 54 miles. An extension to Jackson, the capital of the state, is in progress.

NEW ORLEANS AND NASHVILLE RAIL-ROAD, see Louisiana.

MISSISSIPPI RAIL-ROAD, extending from Natchez through Gallatin and Jackson to Canton, 150 miles long, in progress.

JACKSON AND BRANDON RAIL-ROAD, length 14 miles.

GRAND GULF AND PORT GIBSON RAIL-ROAD. Length, 7.25 miles.

Rail-roads are proposed: From Natchez to Woodville, 41 miles. From Manchester to Benton, 14 miles. From Brandon to Mobile. From Princeton to Deer Creek, 20 miles; and from Columbus to Aberdeen.

Aggregate length of rail-roads in Mississippi, 83.00 miles.

LOUISIANA.

An extensive system of improvement has been lately established in this state, which from its peculiar configuration, is susceptible of great improvement, at a very moderate cost. The whole of the southern half of the state is nearly level, and is intersected in all directions by what are there called *Bayous*, Lakes and Lagoons, which could be readily united and rendered navigable with but a small amount of labour. By connecting the streams which run in the direction of the Gulf coast, not only a navigable communication would be opened between the Mississippi, both above and below New Orleans, but also a channel for the surplus water of that river would be opened, by which the effects of the annual overflow would to a great extent be avoided. In this way, also, it is probable, the millions of acres of land, now worse than useless, one portion of which is constantly, and the other occasionally submerged, could be reclaimed, and thus brought into cultivation ; and what would be of infinite importance, the salubrity of the country materially promoted thereby. Our only purpose, however, is to point out the facilities for internal navigation, which Louisiana possesses in an uncommon degree. Two great outlets from the Mississippi could be created at a comparatively small expense. One by the numerous lakes and the connecting bayous, which lie between the Mississippi river, below Donaldsonville, and Barataria Bay ; and the other by the Iberville and Amite rivers, and Lakes Maurepas, Pontchartrain and Borgne. The latter river especially, has always, from the earliest settlement of the country, been regarded as an important outlet, equally practicable and effectual. On inspecting the map, the reader will see how admirably the streams and lakes of this region are adapted to the construction of a navigable communication from the Mississippi to the Gulf, which may be then approached through the Regolets and Lake Borgne.

The Iberville runs into Amite river, and from their junction, sixteen miles from the Mississippi, the united streams present a fine body of water to the lakes, sufficient for all the purposes of navigation. The depth of the water in Pontchartrain is generally from eighteen to twenty feet. The bay of Manshac was opened some years since by General Wilkinson, wide enough for the passage of boats; but, during the late war, the American commander, apprehending the approach of the British troops through that channel, ordered it to be obstructed, by falling a quantity of cypress trees across it, which presents an obstacle to navigation, until they are removed. It is believed that, by clearing out these obstructions, deepening and widening the bed, constructing small levees for a short distance, and cutting off a small point at the mouth of this estuary, a considerable portion of the waters of that immense river would find an outlet to the gulf through the lakes; which would greatly improve their navigation, by an accumulation of water sufficient to overcome the feeble resistance of the tides, and form a current outward to the gulf of Mexico. By this mode of conducting off the surplus waters of the Mississippi, two great evils would be avoided—the incumbent waters in the river, and the reflux from the swamps, both of which have been found to be detrimental to the planters on its borders.

It is believed by every person, practical or scientific, that the levees cannot be extended further up the Mississippi, without manifest danger to New Orleans and the contiguous country; and every one must be convinced, that they are inferior to artificial sluices or canals, that would convey the superabundant water to the sea by other routes than the river.

There are several points below the Iberville, at which communications might be formed with Lake Pontchartrain, by cutting less than five miles. One at Bonnet Quarre, where it is said that the river, at low water, is ten feet higher than the lake: the greatest elevation of the river, at that point, during the spring floods, is estimated at twenty-three feet: this quantity, expanding over such a surface as the lake, would produce but a slight effect, whilst it would greatly diminish the body, and, consequently, the danger, of the river.

Through the lakes, and behind the islands which stretch along the coast of the gulf, there is a safe interior passage to Mobile bay, a distance of 170 miles, free for vessels of any size that might be employed in that trade, without any impediments except the slight obstructions between the river and lakes. Between the bays of Mobile and Pensacola, a distance of 50 miles, there are but two interruptions to the water communication; a portage from Bon Secours bay to Perdido, of four and a half miles, and a half mile from the latter to the Grand Lagoon, which communicates with Pensacola bay; making an inland navigation for that distance, by cutting five miles only, almost in a direct line, through a level country, and a soil mixed with clay and sand, furnishing every prospect of easy excavation.

Santa Rosa sound makes out from the bay of Pensacola 40 miles, to Choctawhachie bay, of about the same length; from the end of which, a few miles up East river, will reach a point within five miles of the west end of St. Andrew's bay, through a soil and surface presenting no difficulties to the continuation of the work; from that point to the east end of the bay, in a line with the whole route, is about twenty-four miles; from thence, to the Chipola river, at a point near which there is a large, open, natural communication from the Appalachicola, is about two miles. Thus, with the inconsiderable obstructions at the Mississippi, the removal of small impediments at a few points, and the cutting of twelve miles, an inland navigation may be effected of 350 miles, from the Mississippi to the Appalachicola.

From the lowness of its banks, and the fragility of its levees, the Mississippi often bursts its embankments, and overwhelms the farms that cover its bottoms; and it would be idle to say, that such inundations over fields of cotton and sugar are ruinous in the extreme. To diminish this danger in the slightest degree, would be a national benefit far greater than would be commensurate with the cost to be incurred. By clearing out the Iberville, the first stage in the great route of natural canaling is accomplished, which gives the Mississippi an outlet through which much of its surplus volume would pass, into the lakes first, and then into the gulf, without hazard to its borders and with manifest relief to its levees.

It is thus that another eligible mouth is created, where it is so eminently useful; a portion of its waters, too great for its bed and current, drawn off, a surplus ruinous to its settlers, and hazardous to New Orleans. By this work, all the bays and rivers of the gulf would be connected, and thus furnish a safe and easy conveyance for the valuable timber and productions of their borders, to the ports from which they could be most conveniently shipped.

Regarding the several artificial works of Louisiana, as inseparably connected with its natural navigation, and merely as auxiliary parts of a single system, we shall endeavour to arrange the whole into general lines, and thus develop the entire system in all parts, whether of canals, rail-roads, or natural courses.

Assuming New Orleans as a centre, or point of departure, from which the various lines will radiate, when the system shall have been perfected and brought into successful operation, we proceed to a brief notice of those lines, in such order as their courses and location suggest. That towards the north commences with the PONTCHARTRAIN RAIL-ROAD, which extends 4½ miles, to the south shore of Lake Pontchartrain, about a mile east of the Bayou St. John, and the northern terminus of the ORLEANS BANK CANAL, which has been constructed at the enormous cost of $1,000,000, and which, like the rail-road, unites the Mississippi at New Orleans with Lake Pontchartrain, 4¼ miles distant. A short distance north of New Orleans, the Pontchartrain Rail-road is intersected by the NEW ORLEANS AND NASHVILLE RAIL-ROAD, now in progress. This important work is to extend along the shores of Lakes Pontchartrain and Maurepas; across the Pass Manchac into the valley of the Tangipahao, which it will follow to the line which divides Louisiana from the state of Mississippi. Its course through the latter state is not yet determined on, but it will, no doubt, be in the direction of Florence, in Alabama, and thence nearly direct, intersecting the counties of Lawrence, Maury, Williamson and Davidson, to Nashville, in the latter county. The entire distance from New Orleans to Nashville is about 560 miles. From Nashville the line will be continued by another company, or perhaps by the state, to Knoxville, where it will unite with the Highwassee Rail-road, now in progress in east Tennessee, by which, in connection with the proposed rail-road from the Ten-

LOUISIANA. 189

nessee line, via Abingdon and Lynchburg, to Richmond, in Virginia, an uninterrupted line by rail-roads or steam-boats will be effected to the state of Maine. The great outlet towards the east requires but a small expenditure to render it complete.

Nothing is wanting in Louisiana but the completion of the LAKE BORGNE RAIL-ROAD, about 22 miles in length, when an inland navigation may be obtained, as we have already shown, from its termination on the lake, to Pensacola, and thence, with some trifling cutting, to the Appalachicola river. For the northwestern route we have in addition to the Mississippi, the proposed BAYOU SARA RAIL-ROAD, 101 miles in length, to extend from New Orleans, along the left bank of the Mississippi, to St. Francisville, whence the line is continued by the rail-road now in use, from the latter to Woodville, in Mississippi, which will ultimately be extended to Natchez, and thence to Vicksburg.

At Baton Rouge, this line is intersected by the BATON ROUGE AND CLINTON RAIL-ROAD, 20 miles in length; and at Port Hudson, on the Mississippi, a few miles above the former, commences the PORT HUDSON, JACKSON AND CLINTON RAIL-ROAD, 28 miles long.

This will probably be continued to Liberty, in Amite county, where it will join the SPRINGFIELD AND LIBERTY RAIL-ROAD, 30 miles in length.

Opposite to the southern terminus of the Woodville Rail-road at Point Coupee, commences the ATCHAFALAYA RAIL-ROAD, which extends in a south-west direction, 30 miles from that point, to Opelousas, in the parish of the same name. From Morganza, situated on the right bank of the Mississippi, and on the line of the Atchafalaya Rail-road, to Alexandria, on Red river, no improvement has yet been made: the old mode of communication by steam-boats is still maintained. There is a rail-road now in progress from Alexandria to Cheneyville, about 30 miles in length, which may ultimately be extended to Opelousas, and thus complete a line of rail-road to New Orleans in the south, and Vicksburg on the north. Several other works have been executed or are proposed, extending from New Orleans in various directions. Among them a rail-road to Carrollton, and thence to Fayetteville, 11¼ miles, including branches; another of 4½ miles, to a bend on the Mississippi, below New Orleans, called the English Turn; and another, the Orleans Street Railroad, 1½ miles in length, extending to Bayou St. John.

The canals of Louisiana are few in number, and limited in extent. In addition to the Orleans Bank Canal, before mentioned, there is the BARATARIA NAVIGATION, extending from a point 6 miles above New Orleans to Berwick's bay, on the outlet of the Atchafalaya. It consists chiefly of lakes and streams connected by several short canals, amounting in the aggregate to 22 miles in length, the balance of the line, 63 miles, being composed of natural navigation.

CANAL CARONDELET is an inconsiderable, though a very expensive work. It extends from New Orleans to the Bayou St. John, is about 2 miles in length, and is said to have cost $750,000.

LAKE VERET CANAL, extends from the Lafourche to Lake Veret, and is about 8 miles long.

Other unimportant canals exist in the neighbourhood of New Orleans, and in the parishes bordering on the Mississippi. These have been constructed by individuals for private use, and are generally not intended for navigation.

On looking at the map, it will be perceived that nearly all the canals and rail-roads of Louisiana, though apart and detached from each other, are connected in almost every case with navigable streams.

Most of them diverge from the Mississippi banks, and extend into the interior; they are, in fact, mere ramifications or arms of that great river, and as such will contribute their respective quota to swell its already immense trade.

Aggregate length of rail-roads in Louisiana, 97.25 miles.
" canals " 99.25 "

TENNESSEE.

RAIL-ROADS.

LA GRANGE AND MEMPHIS RAIL-ROAD. Commences at Memphis on the Mississippi, and extends to La Grange in Lafayette county, 50 miles in length. This work is designed to connect with the Tuscumbia and Decatur Rail-road of Alabama, and by that road to reach the Alabama, Georgia and South Carolina Rail-roads, or with the Western and Atlantic Rail-road of Georgia.

SOMERVILLE BRANCH, extends from the main line at Moscow, 16 miles, to Somerville. It is now in progress, and will be finished this year, (1840.)

HIGHWASSEE RAIL-ROAD, extends from Knoxville by Calhoun, to the division line of Georgia, where it unites with the Western and Atlantic Rail-road, of Georgia, 98.50 miles in length. There are no grades on this road exceeding 36 feet per mile, and no curve less than 1000 feet radius. An extension of the Highwassee Rail-road from Knoxville along the valley of the Holston, to Blountsville, is contemplated. This, with the proposed South Western Rail-road of Virginia, from Richmond to the Tennessee line, will perfect an entire Rail-road communication between Richmond and the Alabama Rail-roads.

A RAIL-ROAD from Nashville to Knoxville, the connecting link between the New Orleans and Nashville and the Highwassee Rail-roads, is proposed.

NEW ORLEANS AND NASHVILLE RAIL-ROAD. (See Louisiana.)

CINCINNATI, LOUISVILLE AND CHARLESTON RAIL-ROAD. (See South Carolina.)

Aggregate length of Rail-roads in Tennessee, 164.50 miles.

KENTUCKY.

The improvements of this state are under the direction of a "Board of Internal Improvement," consisting of four members, who are charged with the superintendence of all river improvements; and also to a certain extent of canals and rail-roads, undertaken by joint stock companies.

The improvements authorized by the legislature relate exclusively to the water courses, some of which have been improved, and others are now in progress.

Among these are the following—

KENTUCKY RIVER NAVIGATION, extends from Port William, on the Ohio, to the entrance of the North fork, in Estill county, a length of 260 miles. Its course is exceedingly circuitous, the distance between its terminii in a direct line being only 112 miles. The difference of altitude between those points is 216 feet; overcome by 17 locks, 175 by 38 feet, which connect the various pools formed by 17 dams, varying in length from 350 to 500 feet, and from 20 to 25 feet in height; lowest depth 6 feet; estimated cost, $2,297,409.

LICKING RIVER NAVIGATION. This improvement extends from the confluence of the Licking and Ohio rivers, opposite to Cincinnati in Ohio, to West Liberty in Morgan county, a distance, by the course of the river, of 231, and by a right line, 105 miles. 21 locks, 130 by 25 feet; 21 dams, from 200 to 450 feet long, and from 16 to 27 feet high; difference of level, 310 feet; cost $1,826,481.

GREEN RIVER NAVIGATION, extends from the Ohio to the town of Bowlinggreen, in Warren county, 175, and by a straight line, 90 miles. This like the other river improvements, consists of dams, locks and slackwater pools. The locks are each 160

KENTUCKY. 193

feet long and 36 feet wide, with an average lift of 15 feet; the dams are from 300 to 700 feet long. Cost $473,156.

RAIL-ROADS.

LEXINGTON AND OHIO RAIL-ROAD, commences on the Ohio river, near Louisville, and proceeds nearly due east to Middletown; thence turning towards the north-east, it gains the ridge which divides the waters of the Ohio from those of Floyd's Fork of Salt river. Pursuing the ridge, and passing the towns of Brownsboro and Ballardsville, the line descends the valley of Kentucky river, and enters Frankfort in Franklin county. From Frankfort it assumes an east course, which is maintained until it reaches Lexington, where the road terminates. Length 92.75 miles. Minimum radius of curvature, 1000 feet; maximum inclination 30 feet per mile. It descends the valley of Green river, by an inclined plane, 4000 feet in length, and 240 in height. Cost $1,000,000.

PORTAGE RAIL-ROAD, from Bowlinggreen to Barren River, 1.50 miles in length.

CINCINNATI, LOUISVILLE AND CHARLESTON RAIL-ROAD. (See South Carolina.)

Rail-roads are proposed from Hopkinsville to Cumberland river; from Henderson to Nashville; from Russelville to Clarksville; from Louisville to Bushville; from Louisville to Knoxville; from Madisonville to Pond river; from Covington to Latona Springs, and from Falmouth to Lexington.

CANALS.

LOUISVILLE AND POTTSDAM CANAL. The only canal in this state is the important, though short one, along the falls of the Ohio at Louisville. It is about 2½ miles in length, and has four locks, sufficiently capacious to admit steam-boats of the largest class. The canal is fifty feet wide at the surface. Entire lockage of twenty-two feet. Cost about $730,000. With a trifling exception, the entire line is excavated out of compact limestone, to the mean depth of 10 feet; every thing connected with the canal, is of the most substantial kind, and the mechanical execution of the work throughout, affords a commendable specimen of canal making.

Aggregate of Rail-roads in Kentucky, 94.25 miles.
" Canals " 2.50 "

ILLINOIS.

CANALS.

In addition to the Illinois and Michigan Canal, the construction of which was authorized in 1836; the state legislature in the early part of the year 1837, adopted an extensive system of internal improvements, by means of rail-roads and canals, which has since been vigorously prosecuted under the direction of a "Board of Commissioners of Public Works," in conjunction with a "Board of Fund Commissioners," and an "Engineer in Chief," on whom the duty of making the preliminary arrangements devolved. In conformity with the legislative enactments, the necessary surveys and examinations were made, and submitted to the legislature for their consideration, when a plan of improvement on an extensive scale was concerted, and finally adopted.

There is, perhaps, no section of our Union, of an equal extent, so admirably adapted by nature, for the introduction of a system of rail-roads and canals, as the state of Illinois. Its entire surface is remarkable for its exemption from elevations deserving the name of mountains, and with some trifling exceptions in the north, no prominent hills are to be found within its borders. The whole area presents an almost unbroken plane, but slightly inclined towards the Ohio and Mississippi rivers, the common recipient of nearly all its interior streams. Of these the Illinois, though the principal, when viewed in connection with the improvement of the state, forms only a link in a very extensive system, each member of which assumes a degree

ILLINOIS.

of importance, corresponding with its navigable facilities, and the adaptation of those facilities to the purposes of internal improvement.

No rivers of any consequence exist in the north-eastern part of the state, and it is only in the western and southern portions that streams of much magnitude are found flowing from the interior; such being the hydrography of the former section, the question as to the practicability of supplying the deficiency by artificial means, early attracted the attention of the public authorities; and as the Illinois presented the most feasible route for a channel of communication between the waters of the two great American valleys, that river was examined with a view to this object. This stream rises in the northern part of the adjoining state of Indiana, interlocking with the waters of Lake Michigan, about 350 miles from its entrance into the Mississippi, and with the exception of the dividing land, (ridge it cannot be called) its bed is considerable depressed below the surface of Lake Michigan, from which a canal of any dimensions could be supplied with water.

In the early settlement of the country, it is stated by Volney and other travellers, boats frequently passed from the Lake to the Des Plaines, one of the principal branches of the Illinois. The importance of such a connection, the practicability of which was so obvious, could not fail to impress every beholder, but its execution was reserved for the people of the present day.

This work, which is called the
ILLINOIS AND MICHIGAN CANAL, was commenced in 1836, under the authority of the state government. It extends from a point on the south branch of Chicago river, about five and a half miles from Chicago, along the valleys of the Chicago, Des Plaines and Illinois, to the head of steam-boat navigation on the latter river. The main trunk is 96.35 miles in length, to which must be added 5.55 miles of river navigation along the south branch of Chicago river, and a navigable feeder, 4 miles in length, from Fox river; making a total length of 105.90 miles. It is 6 feet deep; 60 feet wide at top water line, and will cost, according to the last estimate, $8,654,337 51.

Although the canal is limited in extent, and free from the usual obstruction of hills and other elevations, the nature of the earth through which it is to pass, is such as to render its execution exceedingly laborious and expensive. One section, about 7 miles in length, requires a cut 18 or 20 feet in mean depth, through an indurated clay, and this is immediately succeeded by another of nearly similar depth through compact limestone, the whole presenting an extent of heavy excavation into solid rock or its equivalent, almost unparalleled in the annals of canal making. At a distance of 30 miles from the lake, the deep cutting terminates, and at the further distance of 6 miles, the canal makes its first descent from the lake level, by 2 locks, of 10 feet lift each. Thence to its entrance into the Illinois, it maintains a nearly uniform descent.

This, in many respects, is one of the most important canals in our country, and like most similar undertakings its advance, though steady, is very slow.

With the exception of some small canals constructed by private companies, the Illinois and Michigan Canal is the only one in the state.

RAIL-ROADS.

The rail-roads as contemplated by the act of 1837, are important both in point of number and extent. One, which may be regarded as the spine, is to extend through the entire length of the state, from which most of the others diverge, and intersect the various counties on both sides of the principal line, which is sometimes called the CENTRAL RAIL-ROAD. The line as proposed, commences at the city of Cairo, at the confluence of the Ohio and Mississippi rivers, and extends nearly due north, through Vandalia, Shelbyville, Decatur and Bloomington, to the western terminus of the Illinois and Michigan canal, and thence, via Savannah to Galena; a distance of 457.50 miles. $3,800,000 have been appropriated for the construction of this road.

ALTON AND MOUNT CARMEL RAIL-ROAD. This road, when completed, will extend entirely across the state, in an east-south-east direction, passing through the towns of Edwardsville, Carlyle, Salem, Fairfield and Albion. At Mount Carmel it unites with the line of the proposed rail-road to Cincinnati. Length 155 miles.

EDWARDSVILLE AND SHAWNEETOWN RAIL-ROAD, 142 miles

ILLINOIS. 197

in length, branches off from the preceding road at Edwardsville, and with a general south-east course, passes through Lebanon, New Nashville, Pinckneyville, Frankfort and Equality, to Shawneetown.

BELLEVILLE RAIL-ROAD, extends from Belleville to Lebanon, on the Edwardsville and Shawneetown Rail-road, 14 miles.

QUINCY AND DANVILLE RAIL-ROAD, extends from Quincy on the Mississippi, through Columbus, Clayton, Mount Sterling, Meredosia, Jacksonville, Springfield, Decatur, Sydney, to Danville, whence it will be prolonged to the state line in the direction of Lafayette in Indiana. Length 220 miles. The section of this road from Meredosia to Jacksonville, 22 miles in length, is now in use.

ALTON AND PARIS RAIL-ROAD, extends from Alton, through Hillsboro, Shelbyville, Charleston, and Paris, to the state line in the direction of Terre Haute. Length 160 miles.

PEORIA AND WARSAW RAIL-ROAD. From Peoria via Canton, Macomb and Carthage to Warsaw; 116 miles in length.

BLOOMINGTON AND PEORIA RAIL-ROAD, from Bloomington to Peoria, via Mackinaw; 40 miles.

PEKIN BRANCH of the preceding, extends from Mackinaw to Pekin. 20 miles in length.

Some of these sections of the above rail-rails, which connect navigable water courses, have been in progress of construction since 1837. Portions of them, to a limited extent, are now finished, and others nearly so, but from the want of connection, they are, with a few exceptions, almost valueless. As a system they can scarcely be regarded as commenced, as those actually finished and in use, form a very inconsiderable part of the entire system, as contemplated by the legislature.

A Rail-road, 6 miles in length, from Illinois at the Ferry opposite St. Louis, to coal mine Bluffs, is now in operation.

The sum of $11,315,099, has been appropriated to the construction of the above rail-roads, and $400,000 for the improvement of the great Wabash, Illinois, Rock, Kaskaskia, and little Wabash rivers. Only a part of these sums has been expended. Rail-roads are proposed from Alton to Springfield; from Jacksonville to Augusta, 22 miles; from Chicago to Des Plaines, 12 miles; from Naples to Jacksonville, 22 miles; from Alton to Erie; from Belleville to the Mississippi; from Galena to Chicago; from Springfield to Carrolton, and thence to the Mississippi; from Waverly to Grand Prairie; from Lynville to Jacksonville.

Aggregate length of Canals in Illinois, 105.00 miles.
Aggregate length of Rail-roads in Illinois, 26.00 "

INDIANA.

Like the neighbouring state of Illinois, Indiana has manifested a most zealous regard to the cause of internal improvement. There are at the present time, in progress, under the authority of the state, public works, the cost of which, when finished, is estimated at twenty-one millions of dollars. These works, it is probable, will be ultimately united to those of Illinois on the west and Ohio on the east, and thus form a continuous system, both in point of extent and utility, unparalleled either in this or any other country. The aggregate length of canals is 840, and of rail-roads, 90 miles; some of which are completed, and others are now in active progress. The whole expenditure in 1838 was about $1,200,000; and the entire amount of contracts, made since the 1st of March, 1836, when active operations commenced, exceed five millions of dollars.

CANALS.

Wabash and Erie Canal. This important canal which unites the great Mississippi valley with the St. Lawrence Basin, is the joint work of the states of Ohio and Indiana, 87.27 miles of it being in the former, and 99.73, in the latter state: whole length 187 miles. It is proposed to extend the line to Terre Haute and to improve the navigation of the Wabash, at the Great Rapids, and thus open an uninterrupted water communi-

INDIANA. 199

cation from Lake Erie to the Ohio river. A farther extension from Terre Haute to the Central Canal in Greene county, is also proposed. In its passage from Lake Erie, the canal is intersected by the Miami Canal of Ohio; the proposed Northern Canal, and the Central Canal of Indiana. Its course is along the valley of the Maumee, in the north-western part of Ohio and north-eastern part of Indiana; over the dividing ridge between that river and the Wabash; along the right or north bank of the latter; crosses at Tiptonsport and proceeds along the left bank of the Wabash, to Lafayette in Tippecanoe county. 355,200 acres of land have been appropriated by Congress in aid of this important undertaking.

The district for which this canal will form the main channel of trade, may be described as extending from the state line as far down the Wabash as the Grand Rapids, a distance of 300 miles. The boundaries of the district on the south and southeast may be defined by a line pursuing generally the valley of the west fork of White river to the east line of the state, embracing nearly one-third of the surface between the Wabash and the Ohio river; and on the north and west by a line diverging from the Grand Rapids, and extending about one-third the distance to the Illinois river on the west, and Lake Michigan on the north. The limits of this district, it will be perceived, are marked out with due reference to the influence of the Ohio navigation on the south, and of the Illinois river and Lake Michigan on the west and north, as rival channals of commerce. The district thus described contains a surface equal to thirty-eight counties in Indiana, and nearly nine counties in Illinois, including an aggregate area of 22,000 square miles. The population of this district, in 1839, as nearly as can be estimated from official reports of the state officers, was about 270,000, averaging $12\frac{1}{2}$ persons per square mile. Allowing for the increase of the population at the usual rates in similar districts, the average by 1841—the time when it is supposed navigation might be opened to the lake—will be increased to about 15 per square mile.

The Ohio division of the Wabash and Erie Canal is 87.27 miles in length, and is estimated to cost, by the Ohio Board of Public Works, in their annual report of December 30th, 1839, $2,000,000.

Indiana will send down through the Ohio portion to Lake Erie, at least 100,000 tons of freight annually. By the terms of the compact between the states, the state of Ohio is authorised to levy upon the commerce of Indiana, the same toll that is charged on her other principal canals. These rates average two cents per ton per mile, which, multiplied by 87.27, the length of the canal in the state of Ohio, will give $1 75 per ton. This multiplied by 100,000, the number of tons passing through in a season, will give a revenue to Ohio from the commerce of Indiana alone, of $175,000. Deduct from this $35,000 for repairs, collection of tolls, &c. equal to $400 a mile, and the state of Ohio will realize an income of $140,000 from the business of her sister state. This sum (the $140,000) is just 7 per cent. on the whole cost of the work as stated above.

CENTRAL CANAL, intersects the south bank of the Wabash and Erie Canal at two points, one near Peru in Miami county, and the other at Wabash in Wabash county. Proceeding towards the south-east, these sections unite near Marion in Grant county, whence the canal advances due south to Andersontown in Madison county, where it enters the valley of the West Fork of White river. From Andersontown its course is due west to Strawtown in Hamilton county, where it turns towards the south-west, passes through Noblesville, and after a course of 60 or 80 miles, enters Indianapolis, the capital of the state. Resuming its south-western direction, it follows the east declivity of the North Fork, through the counties of Marion, Morgan, Owen, Greene, and Daviess, and the towns of Martinsville, Spencer, Bloomfield and Maysville, to Petersburg in Pike county. There the canal leaves the bank of White river, and after a meandering course through Gibson and Vanderburg counties, terminates at Evansville, on the Ohio river. Entire length 290 miles. Several divisions of this work are now in course of construction. Estimated cost $3,500,000. A branch to extend from Andersontown to Muncietown, is proposed.

WHITEWATER CANAL, extends from Lawrenceburg in Dearborn county, on the Ohio river, ascends the valley of Whitewater, through Hamilton county in Ohio, and Franklin, Lafayette and Wayne counties of Indiana, to Cambridge on the National Road, in the last named county. A considerable

INDIANA.

portion of this work is now in use. The section from Lawrenceburg to Brookville, 30 miles, was opened for navigation in 1839; whole length, 76 miles. Estimated cost $1,400,000. An extension to the Muncietown branch of the Central Canal is proposed.

TERRE HAUTE AND EEL RIVER CANAL, is an extension of the Wabash and Erie Canal from its southern terminus at Terre Haute, via the valley of Eel river, to the Central Canal in Greene county. Length 40.50 miles. Estimated cost $629,631.

A canal from Wayne on the Maumee to Michigan City in the north-west angle of the state, and thence to the Illinois boundary, is proposed, but not yet definitively located.

RAIL-ROADS.

MADISON AND INDIANAPOLIS RAIL-ROAD, commences at Madison, in Jefferson county, on the Ohio river, and proceeds through the counties of Jennings, Bartholomew, Johnson and Marion; and the towns of Vernon, Columbus, Franklin, &c. to Indianapolis, the capital of the state. Length about 95 miles. It is completed from Madison to Vernon, a distance of 25 miles.

LAFAYETTE AND MICHIGAN RAIL-ROAD, 106 miles in length. This road is designed as a continuation of the line of thoroughfare from the Ohio river to Lake Michigan, of which the Madison and Indianapolis Rail-road and the M'Adamized road from the latter to Lafayette, form a part.

In addition to the above which are state works, other railroads are proposed, by joint stock companies, to extend from Jeffersonville on the Ohio, opposite Louisville, via Vienna, Rockford and Columbus, to Indianapolis, 108 miles; a small part of which is now under contract: from Terre Haute to Evansville, via Princeton and Vincennes: from Lafayette to the western line of the state, in the direction of Danville in Illinois, where it will intersect the Quincy and Danville Railroad of that state: from New Albany, through Clark, Scott, Jackson and Bartholomew counties, to Columbus, where it will join the Madison and Indianapolis Rail-road: from the eastern to the western boundary of the state, near its northern border:

from Lawrenceburg to Indianapolis, via Greensburg and Shelbyville, and from Michigan City to Laporte.

Aggregate length of canals in Indiana, 217.00 miles.
"   "   rail-roads   "   95.00   "

OHIO.

Our remarks on the physical structure of the state of Illinois, apply with nearly equal force to that of Ohio. For the construction of canals and rail-roads, the entire region comprehended by the states of Illinois, Indiana, Ohio and Michigan, presents fewer impediments to the construction of such works, than any other with which we are acquainted. It requires but a slight examination of the natural features of this section of the United States to sustain our position. This portion of the Ohio valley, it is obvious, once composed an extensive inclined plane, into which the beds of the streams have been formed by the slow but equally sure process of abrasion.

The hills are generally found near the rivers or large creeks, and parallel to them on each side, having between them the alluvial valley, through which the stream meanders, usually near the middle, but sometimes washes the foot of either hill alternately. Perhaps the best idea of the topography of this state may be obtained by conceiving the state to be one vast elevated plain, near the centre of which the streams rise, and in their course wearing down a bed or valley, whose depth is in proportion to their size, or formation of the earth over which they flow. So that the hills, with some few exceptions, are nothing more or less than cliffs or banks, made by the action of the streams; and although these cliffs or banks on the rivers or larger creeks, approach the size of mountains, yet their tops are generally level, being the remains of the ancient plain. In the eastern part of the state, some few hills are found in sharp ridges, similar to those in the eastern states. The bases of the hills are generally composed of limestone, free or sandstone, slate and gravel, admixed with mineral coal, ochre, &c. The entire valley of Ohio rests on horizontal strata, belonging to that formation called by geologists floetz or secondary. Near Pittsburg, the rocks are so nearly parallel with the horizon, as to scarce admit a current from the deep perforations of the coal mines. These mines are opened along the sides of the hills, and extend inwards on a level with the horizon, and about 320 feet above the lower surface of the adjacent rivers. The circumstance most conclusive of the fact, that the hills and vallies of this region were formed by abrasion, is the uniformity of elevation, and similarity of material, of corresponding strata, on the opposing banks of the streams : phenomena, however, every where visible, in Ohio valley, where the nature of the country will admit accurate observation.

The Ohio valley is subdivided by the Ohio river into two unequal sections, leaving on the right or N. W. side, 80,000, and on the left or S. E. side, 116,000 square miles; the Ohio river flowing in a deep ravine, and forming a common recipient for the water poured down from both slopes. The length of the Ohio ravine, in a direct line from the city of Pittsburgh to the Mississippi river, is 548, but by the meanders of the stream 948 miles.

The peculiar features of this river, and its immediate banks, have led to most of the gross misrepresentations respecting the valley in general. The low water surface of the Monongahela at Brownsville, is 850, and at Pittsburg, 830 feet above the tides in Potomac river at Washington city. The apex of the hills around Pittsburg are within a small fraction of 460 feet above low water level in the rivers in the same vicinity. These elements give us 830 to be added to 460, or 1290 feet, as the extreme elevation of the hills near Pittsburg. The data being in great part drawn from actual admeasurement may be considered as correct, and combining the result with the hypothesis of the whole valley being once an inclined and unbroken plain, we are led to the conclusion that about 1306 feet in round numbers was once the general elevation of that plain, where the Monongahela and Allegany now form the Ohio. The plain must have risen considerably higher towards the Allegany system, and towards lake Erie, and declined slowly towards the Mississippi and Illinois rivers; and such depression, though more gradual, must have continued until the land sunk under the Gulf of Mexico.

The elevation of surface at the central junction of the Ohio and Mississippi, has not been determined with the same precision as has been done respecting that near Pittsburg, but may be estimated with considerable accuracy from the length of the Mississippi below the mouth of the Ohio, which is very nearly 1100 miles. If we allow $3\frac{1}{2}$ inches fall to each mile, we shall have 3850 inches, equal to 321 feet within a very small fraction, for the height of the country at the junction of Ohio and Mississippi rivers. Deducting 321 from 830, would leave 509, as the fall in the Ohio; but this sum exceeds the real depression of that stream. A very considerably greater fall exists from Pittsburg into Chesapeake bay, than into the Gulf of Mexico, a seeming anomaly explicable from the simplest laws of hydrostatics. The Gulf of Mexico is a real reservoir, supplied by the Gulf stream, and evidently elevated above any other part of either ocean which laves the coast of America. The Gulf stream flows from the Gulf of Mexico into the Atlantic ocean with great velocity, and the current, though continually lessening, is continued from the Bahama channel to the coasts of Europe and Africa, by a curve of upwards of six thousand miles; but if we restrict our view to the higher part of the tropic current, or that from Cuba to Chesapeake bay, or about 1000 miles, the velocity of the stream must demand at least an inch fall per mile, or 83 feet. If this hypothesis is correctly formed from existing data, then is the surface of Chesapeake bay 83 feet depressed below that of the Gulf of Mexico, and of course the fall of water from Pittsburg into the latter recipient only 747 feet.

It is a fair induction from what has been stated, that the valley of Ohio is composed of an inclined plane, furrowed by the deep channels of the rivers, and chequered by hills and alluvial flats, the whole resting on a floetz or secondary formation. In some parts of the basin, particularly in the state of Kentucky, the rivers flow in chasms rather than valleys, in the true meaning of the latter term. The two opposing slopes present some curious contrasts. Though most extensive, the south-eastern slope has no considerable remains of the ancient plain; the north-western slope on the contrary, contains in the central parts of Ohio, Indiana and Illinois, large tracts marking unequivocally the primitive state of the valley. The confluents of Ohio, which flow from the Allegany mountains, are precipitous torrents from their sources, and, as has been already noted, pursue their courses in deep channels; whilst those streams which derive their fountains from the north-western slope, rise on a continuous plain, in some places morass, sluggish towards their sources, but gaining velocity as they approach the Ohio.

The principal confluents of Ohio from the south-eastern slope, are the Monongahela, Little Kanawha, Great Guyandot, Sandy, Licking, Kentucky, Greene, Cumberland and Tennessee. Those flowing from the north-west are the Allegany, Beaver, Muskingum, Hockhocking, Sciota, Miami and Wabash. Of these streams, the Allegany and Monongahela are the constituents of Ohio; the former rising in Pennsylvania and New York, and fed by numerous branches, pursues a general course of S. a little W. 200 miles, but with a very circuitous channel, and unites with the Monongahela at Pittsburg. The latter rises in Virginia at N. lat. 38°, by two branches, the Monongahela and Cheat; draining Pocahontas, Lewis, Randolph, Preston, Harrison and Monongalia counties, unites immediately within the southern boundary of Pennsylvania, and continuing by a general course nearly north, joins the Allegany, and forms the Ohio, after a comparative course of 150 miles, but perhaps 200 by the windings of the streams.

The sources of the Allegany are the extreme north-eastern tributaries of the Mississippi basin, and flow from the highest part of the Ohio valley. Westward from the valley of the Allegany, that of the Beaver exhibits the commencement of the central plain which divides the basins of the Mississippi and St. Lawrence. This plain stretches westward, and widening in extent over the states of Ohio, Indiana, and Illinois, reaches the Mississippi river. In its natural state, the valley of Ohio was generally covered with a very dense forest, but the central plain presented an exception. As far east as the sources of Muskingum, commenced open savannahs, covered with grass, and devoid of timber. Similar to the plain itself, those savannahs or prairies expanded to the westward, and on the waters of Illinois opened into immense natural meadows, generally known under the denomination of prairies.

It has been shown that Pittsburg was elevated 747 feet above the surface of the Gulf of Mexico. Lake Erie has been found 565, and Pittsburg 830 feet above tide water in the Atlantic bays of Chesapeake, Delaware, Hudson and St. Lawrence; consequently Pittsburg is elevated 265 feet above lake Erie; the intermediate distance in a direct line, 106 miles. Therefore, if a channel could be opened from the level of Ohio at Pittsburg, as deep as the bottom of that river, and carried into lake Erie, the water of Allegany and Monongahela, in place of flowing toward the Gulf of Mexico, would rush into lake Erie with a velocity of 265 feet in 105 miles, or upwards of $2\frac{1}{2}$ feet per mile.

A due attention to these mathematically established facts, will enable the reader to comprehend the real structure of the higher part of the valley of Ohio. Nothing indeed but real admeasurement could render credible, that the Allegany river should have part of its source within five miles from the margin of lake Erie, and after winding from thence 200 miles, receive a large outhern branch, and be still 265 feet above the surface of the lake. In fact, the Ohio does not sink to the level of lake Erie before having flowed as low down as the vicinity of Marietta, and the mouth of Muskingum.

Another feature in the Ohio valley, is in a peculiar manner interesting; that is, the real slope of its surface. At a first glance upon the map, it would be naturally supposed that from the sources of Allegany and Monongahela, the plain would depress towards the final recipient, the Mississippi; but such is however, not the fact. It is well known that, during the continuance of spring floods, loaded boats of considerable size can be navigated from the rapids of Ohio at Louisville, by the Ohio, Mississippi and Illinois rivers into lake Michigan, and to the head of Niagara falls, without meeting a single rapid; whilst the direct line between the two extremes passes over an elevated ridge.

It has been found that the surface of the Mississippi at the mouth of the Ohio is elevated 321 feet above the Gulf of Mexico. Lake Michigan is about 35 feet higher than lake Erie, or

600 feet above the Atlantic tides. In most parts of its course, Illinois river has much more the aspect of a winding canal than that of a river, in the true meaning of the latter term, there being only 279 feet fall from the level of lake Michigan to the mouth of the Ohio, in a distance of 520 miles, following the meanders of the rivers, or a small fraction above six inches per mile. These elements demonstrate that no part of Illinois river is as high as the bottom of Ohio at the mouth of Sciota, and only near the vicinity of Cincinnati do the two rivers come on the same level ; that the great original plain sloped from the Allegany system towards the Illinois river and Michigan lake ; and that the Ohio traverses the declination of the intermediate space obliquely.

As a navigable section of the United States, the valley of Ohio has some other peculiar features. The Ohio itself, and its principal source, the Allegany, are in a striking manner gentle as respects current, and from Olean in Cataraugus county, New York, to the Mississippi, over a distance of 1158 miles, following the streams, at a moderately high flood, meets, except the rapids at Louisville, with not a single serious natural impediment. The Monongahela, more impetuous than the Allegany, is yet navigable, without falls or rapids, by both branches, far into Virginia. Descending the valley, the two largest confluents from the south-east, the great Kanawha and Tennessee, rise, by interlocking sources, in Ashe county, North Carolina, and flowing in directly opposite courses, each reaches its recipient, the Ohio, by an immense curve, which taken together, sweeps rounds the rivers of Kentucky, and some of those of Virginia and Tennessee. Rising on the highest Allegany table land of the United States, at an elevation of at least 2000 feet, the currents of both Tennessee and Kanawha are extremely rapid ; the latter impeded by falls, and the former by rapids at the Muscle Shoals, but both navigable downwards from near their sources. Though scarcely reaching the spurs of the Allegany system, the rivers of Kentucky, though generally without falls or rapids, have very strong currents, arising from the great descent of their common slope.

On the north-west side of the valley, though from a different structure, the rivers are also extremely rapid. Rising on a table land, from 300 to 1000 feet above their mouths, and in no instance having a direct course of 300 miles, the streams, though falling gradually, are real torrents. The Big Beaver, Muskingum and Hockhocking, have direct falls; but the Sciota, Miami, and Wabash, though excessively rapid, have neither falls no cataracts to impede navigation.

From these considerations, it is manifest that the Ohio valley may be regarded as a great plain inclining from the Allegany, system to the N. W., and obliquely and deeply cut by the Ohio and its numerous confluents, into chasms from 460 feet, to nearly the level of the streams. In the higher part of the valley, when on the riv s, the banks, with the exception of comparatively narrow flats, near the margins, rise by bold acclivities into hills which have a mountainous aspect. This boldness of outline imperceptibly softens descending the Ohio, and, approaching the Mississippi, a monotonous ring of level woodland bounds the horizon. Ascending the rivers of the south-east slope, the scenery becomes more and more rugged, until terminating in the ridges of the Allegany chains; on the contrary, if the rivers of the north-west slope are ascended, we find the landscape broken and varied near the Ohio, but around their sources flat and monotonous, and hence peculiarly fitted for the introduction of internal improvements. In addition to the evidence we have adduced in support of our position, the remarkable regularity of profile of the leading canals of this state, the summit of the Ohio and Erie Canal being only 405 feet above Lake Erie, and 499 above the Ohio at Portsmouth, may be cited.

These advantages were not overlooked by the sagacious and enterprising people of Ohio, who notwithstanding the recent settlement of the country, and the danger of incurring heavy responsibilities, boldly entered upon the great work for which nature had prepared the way.

When the state of Ohio adopted her great system of internal improvement in 1825, she had no other means under her control but direct taxation, and only a few thousand dollars then in the treasury could be appropriated to the purposes of internal improvement. Trusting, in the outset, to an untried credit, and ultimately to her latent resourses, a rich soil, and hardy, industrious cultivators, the legislature authorized the Commissioners of the Canal Fund to borrow on the credit of the state, at an interest not exceeding six per cent. such sums as were estimated to be sufficient to complete the canals. Accordingly, arrangements, highly favourable to the state were made, chiefly with eastern capitalists, and the Manhattan Bank, in the city of New York, became the agent of the state; and as such, made the transfers and paid the interest to the stockholders. Thus relieved from pecuniary embarrassment, the great works were steadily prosecuted under the management of a judicious Board of Commissioners, aided by practical engineers; and were brought to a successful conclusion, with but little aid from the higher attainments of science, or the splendid theories of modern times. In the construction of the canals, little was done for show—much for solidity and convenience. The conductors of these works appear to have spent but a small portion of their time in calculating the form and pressure of arches; they knew that their quarries would furnish materials of great strength and magnitude; they could grasp and bind the catenarian curve between the extrados and the intrados, and the exterior beauty of their structure was quite a secondary consideration. The character of the work bears ample testimony to the skill and fidelity with which it has been accomplished; and affords a commendable example, both in design and execution, that might have been advantageously followed by some of the "eminent engineers" of other quarters, whose inflated and egotistical reports present a striking contrast with the unpretending and business-like statements which we have examined in relation to the Ohio works.

CANALS.

The chief canal in this state is the

OHIO AND ERIE CANAL, extending from Portsmouth, at the intersection of the Scioto with the Ohio river, to Cleveland, on Lake Erie, has been in operation several years. After leaving Portsmouth the canal crosses the Scioto, and pursues a course nearly due north, along the right bank of that river; passes Chillicothe and enters Circleville, where it re-crosses the Scioto, and continues its route along the valley of the Scioto, to its intersection with the Columbus Feeder. Here the canal suddenly turns and pursues an eastern direction through the towns of Hebron, Newark, Irville, and Cosochton, into the valley of Tuscarawas river, which it follows to the summit, after passing through Newcomerstown, Salem, Schoenbrun, New Philadelphia, Bolivar, Massillon, Clinton, &c. On leaving Akron, at the Portage summit, the canal descends the valley of the Cuyahoga, which it follows, and terminates at Cleveland. Length from Portsmouth to Cleveland 307 miles; summit level 499 feet above the Ohio at Portsmouth; 405 feet above Lake Erie, and 973 feet above the Atlantic Ocean; general course northeast; 40 feet wide; 4 deep; 152 locks; lockage 1,185 feet; commenced in 1825; completed in 1832; cost $5,000,000.

COLUMBUS BRANCH, along the left bank of the Scioto from Columbus, and unites with the main canal in the north part of Pickaway county. Length 10 miles. Its structure is similar to that of the main trunk.

LANCASTER BRANCH, extends to Lancaster, 9 miles.

ATHENS EXTENSION, OR HOCKING CANAL, is a prolongation of the Lancaster Branch. Its course is nearly south-east through the counties of Fairfield, Hocking and Athens, to the town of Athens; about 50 miles in length.

ZANESVILLE BRANCH, extends from the main line, along the west branch of the Muskingum, to the town of Zanesville. Length 14 miles.

WALHONDING BRANCH, descends the valley of the Walhonding, and intersects the main trunk at Caldersburg, opposite to Coshocton. Length 23 miles.

GRANVILLE BRANCH, six miles in length.

EASTPORT BRANCH, four miles.

DRESDEN BRANCH, two miles in length. By this work a navigable communication is opened between the Ohio and Erie Canal near Dresden with the Ohio river at Marietta, a distance of 100 miles. It is effected by means of dams and locks, erected in the Muskingum river.

WHITE WATER CANAL. See Indiana.

MIAMI CANAL, now in operation, extends from Cincinnati, on the Ohio river, along the ravines of Mill Creek, the Great Miami, and Auglaize, to Defiance on the Maumee; passing in its course the towns of Springfield, Hamilton, Middletown, Franklin, Miamisburg, Dayton, Troy, Hardin, St. Mary, Paulding, &c. Length from Cincinnati to Dayton, 68 miles; and thence to Defiance, where it unites with the Wabash and Erie Canal, 110; total length 178 miles; general course nearly north; summit level at Dayton 175 feet above the Ohio at Cincinnati, and 606 above the Atlantic; as deduced from a continued series of levelling operations from tide water on the Hudson to Lake Erie, and thence to the several points just mentioned. The dimensions are the same as those of the Ohio and Erie Canal. Completed in 1830; cost $3,750,000.

WARREN CANAL, a branch of the preceding, 20 miles in length, extends in a south-east direction from near Middletown to Lebanon, the seat of justice of Warren county.

SANDY AND BEAVER CANAL, connects the Ohio State Canal with the Ohio river and the Pennsylvania State Canal, at Pittsburg. It leaves the Ohio Canal at Bolivar and passes through the valley of the Sandy branch of Tuscarawas river, in Stark county, over the dividing ridge in Columbiana, and enters Pennsylvania near the mouth of Little Beaver Creek; about 30 miles below Pittsburg. Length 76 miles; cost $1,500,000.

MAHONING CANAL, like the Sandy and Beaver Canal, unites the Pennsylvania and Ohio canals. It leaves the latter at Akron, in Portage county, Ohio, pursues the left bank of Cuyahoga Creek, through the town of Ravenna, and thence into and along the valley of Mahoning river to its confluence with the Beaver river, where it meets the Beaver division of the Pennsylvania Canal, near the town of New Castle, in Mercer county. Length 77 miles in Ohio, 8 in Pennsylvania; cost $764,372.

MILAN CANAL, opens a communication for steam-boats from the head of navigation on the Huron to Milan, a distance of 3 miles, to which the Lake Erie steamers can now ascend.

Canals are proposed from Clinton to Chippeway; from Belleville to Bolivar; from Franklin to New Lisbon; from Mount Vernon to the confluence of Mohiccon and Vernon rivers; from Lower Sandusky to the mouth of Tyemochte Creek; from Cincinnati to Harrison; from Columbus to Delaware; from the mouth of the Chagrino to Holmes's Mill; from Cincinnati to near Harrison; from Columbus to Delaware; and some others.

RAIL-ROADS.

LITTLE MIAMI RAIL-ROAD. The line, as proposed, commences near the centre of High street, in the city of Cincinnati, crosses Crawfish Creek, and passing up the valley of Duck Creek, enters the valley of the Little Miami. Proceeding along the Miami to its junction with Highland Creek, where the line diverges from the former, ascends the ravine of that creek, and re-enters the valley of the Little Miami, opposite to Milford; thence through Lockport and over Obannan and Todd's Creeks, to Waynesville, where the line attains an elevation of 305 feet above the Ohio. From Waynesville it proceeds nearly north, crosses the valley of Glady Creek, near its mouth, and enters the town of Xenia, 498 feet above the Ohio; thence through Clifton to Springfield, which is 534 feet above the Ohio. Length 85.50 miles; estimated cost $877,663 74. Single track with turn-outs and side lines; maximum inclination 40 feet per mile.

For more than half the distance, the grades are under 10 feet per mile. From Xenia to Cincinnati the line has a descending grade nearly the whole distance.

MAD RIVER AND SANDUSKY CITY RAIL-ROAD, is to extend from Dayton, in Montgomery, to Sandusky city, in Erie county. It will pass through Clarke, Champaign, Logan, Hardin, Crawford, Seneca and Sandusky counties, and the towns of Springfield, Urbana, Bellefontaine, Tyemochte, Tiffin, &c. The road is now finished and in use from the latter to Sandusky city, a distance of 30 miles. This road, which may be regarded as an extension of the Little Miami Rail-road, is about 155 miles in length, but from Springfield, where it meets that road, to Sandusky, it is only 128 miles. Its construction is similar to the other rail-roads of that country: wooden sleepers and iron rails. Estimated cost per mile $11,000.

SANDUSKY CITY AND MONROEVILLE RAIL-ROAD, now in progress, extends from the former in Erie county, to the latter in Huron county, 18 miles.

OHIO RAIL-ROAD, extends from the town of Manhattan, on the Maumee river, to Sandusky, and thence to the Mad River Rail-road. Length about 40 miles. A large portion of this road is laid in the Black swamp, in which piles were driven to sustain the road bed.

OHIO. 213

TOLEDO AND KALAMAZOO RAIL-ROAD, see Michigan.

Rail-roads are proposed from Norwalk to Huron, 12 miles; from Akron to Perrysburg; from Ashtabula to Liverpool; from Bridgeport to Sandusky city; from Chillicothe to Cincinnati; from Circleville to Cincinnati; from Cleveland, via Columbus, to Cincinnati; from Cleveland to the Pennsylvania line, in the direction of Pittsburg, Pa.; from Cleveland to Warren; from Columbus to Lower Sandusky; from Columbus to Springfield; from Columbus to Big Spring; from Conneaut to the Pennsylvania line; from Cleveland to Franklin; from Cayahuga Falls to Cleveland; from Wayne to Piqua; from Mansfield to New Haven; from Melmore to Republic; from Zanesville to the Ohio, in Belmont county; from Newark to Mount Vernon; from New Haven to Monroeville; from Akron to Defiance; from the state line, in Ashtabula, to the Miami river and Wabash and Erie Canal; from Stillwater to the mouth of the Maumee; from Toledo to Sandusky city; from Urbana to Columbus; from Vermillion to Birmingham; from Wellsville to Fairport; from Richmond to Miami; from Port Clinton to Lower Sandusky; from Franklin to Wilmington, via Springboro; from Erie to Ohio; from Columbus to Sandusky; from Cincinnati to Indianapolis; from Milan to Newark; from Milan, Columbus, Chillicothe, to Lebanon.

From Bellefontaine, in Logan county, to Perrysburg, in Wood; capital $400,000. From Charleston, in Lorain county, to Ashland, in Richland, via Oberlin; capital $300,000. From Charleston to Elyria; capital $30,000. From Carrollton to Lodi; capital $100,000. From Lima to Shanesville, via Auglaize; capital $100,000. From Massillon to the Ohio river; capital $1,200,000. From Sandusky city to Maumee, in Lucas county, to unite with the Toledo Rail-road; capital $100,000. From Norwalk to Huron; from the Ohio river, in Columbiana county, to the Indiana state line; from Venice to Bellevue; capital $25,000. From the mouth of Vermillion river, in Huron county, to Ashland, in Richland county; capital $300,000; and from Wellsville to Steubenville; capital $500,000; from Wellsville to Fairport.

Aggregate length of canals in Ohio, 779.00 miles.
        "   rail-roads   "   70.00   "

ARKANSAS.

NOTHING has yet been done in this state in the way of canals or rail-roads.

MISSOURI.

THIS state has not as yet accomplished any work of internal improvement. Several works, mostly rail-roads, are projected; and in 1839, the legislature authorised the formation of a Board of Internal Improvement, whose duties were defined by the act. They relate chiefly to the improvement of the natural navigation throughout the state, and to a survey for a rail-road from St. Louis to the Iron Mountain in Madison county.

There are several other rail-road projects on foot: one from St. Louis to St. Charles, and thence westward, through the counties bordering on the north bank of the Missouri; one from the same point and in the same direction, through the southern counties; and one from the town of Louisiana to Columbia, and thence by the one first mentioned above, to Rocheport on the Missouri.

MICHIGAN.

RAIL-ROADS.

IN compliance with an act of the legislature of this state, passed in 1836, a Board of Commissioners was formed with authority to effect a loan of $5,000,000 for purposes of internal improvement. A system of improvement by canals and rail-roads was soon after adopted; and some of the works therein contemplated, were immediately commenced. These improvements consist of three extensive rail-roads and three canals, together with some slack water navigation along the St. Joseph, Kalamazoo and Grand rivers. With regard to the execution of these important lines of thoroughfare, not much has yet been done. The rail-road from Detroit to St. Joseph on Lake Michigan, is progressing slowly towards the west, and is completed and in successful operation from Detroit to Ann Arbor, a distance of 44 miles. This is called the

CENTRAL RAIL-ROAD: it commences at Detroit and pursues a nearly direct west-south-west course to Ypsilanti, where it deflects towards the north-west and advances to Ann Arbor. Thence the line as located is continued by a general western course, through Washtenaw, Jackson, Calhoun, Kalamazoo, Van Buren and Berrien counties, to St. Joseph, on the eastern shore of Lake Michigan. This road, it will be seen, connects Lake Michigan with Lakes St. Clair and Erie, and forms a very important part of the great western thoroughfare for the northern and western states. Already many thriving towns and settlements have been established along the line in anticipation of its ultimate completion. The length of this road, as located, is about 194 miles, and estimated cost of construction $1,928,-195, or nearly $10,000 per mile. The next state work is the

SOUTHERN RAIL-ROAD: extending from the river Raisin, a

short distance below the town of Monroe, to the new village of Buffalo, in the extreme south-west angle of the state and on Lake Michigan, about 30 miles south-west from St. Joseph, the western terminus of the Central Rail-road. The road attains its greatest altitude (631 feet) in the county of Hillsdale, a few miles east from Jonesville, whence it descends to Lake Michigan 14 feet above Lake Erie. The plan of construction of this road, like all the other state rail-roads, is of wood with iron rails. Maximum inclination 40 feet per mile, and the minimum radius of curvature 2000 feet. Average inclination 15.75 feet per mile.

The road as proposed will traverse the counties of Monroe, Lenawee, Hillsdale, Branch, St. Joseph, Cass and Berrien, and terminate at New Buffalo. In its course, the line passes through or near the towns of Adrian, where it is intersected by the Erie and Kalamazoo Rail-road from Toledo, Hillsdale, Branch, Centrevile, Adamsville, Edwardsville and Bertrand. These towns are all situated in the southern parts of the state, for the accommodation of which the road now under consideration, was designed. Length 183 miles. Estimated cost $1,496,376 or $8,176 92 per mile.

HAVRE BRANCH. This road was commenced by a joint stock company in 1836. It has since been transferred to the state. It unites the Southern Rail-road with the Erie and Kalamazoo Rail-road at the town of Havre. Length 13 miles. Estimated cost $82,043, or about $6,360 per mile.

NORTHERN RAIL-ROAD, about 201 miles in length, the last of the state works of this description, commences at Port Huron, on St. Clair river or strait, and near the outlet of Lake Huron; passes through the counties of St. Clair, Lapeer, Genesee, Shiawassee, Clinton, Ionia, Kent and Ottawa, and intersects the eastern shore of Lake Michigan at Grand Haven. The following towns are on the line of this road: Lapeer, Leroy, Owasso, Lyons, Saranac, Ada and Grandville. The summit level, in Lapeer county, is 300 feet above Lake St. Clair; maximum inclination 30 feet per mile; minimum radius of curvature 5000 feet. Estimated cost $1,310,361, or about $6,504 per mile, exclusive of buildings, apparatus, &c. About seven miles west of Grandville, the line, where it curves towards the north-west, is intersected by the Port Sheldon

MICHIGAN. 217

Rail-road, about six miles in length, now in progress under the direction of a joint stock company. Among the works now in progress by private companies,

THE ERIE AND KALAMAZOO RAIL-ROAD is one of the most important in the state. It is in fact an extension of some of the leading roads of Ohio, and as such, must partake largely of the travel through the middle and northern counties of that state. The finished section, 33 miles in length, extends from Toledo in Lucas county, Ohio, at the northern termination of the Ohio Rail-road from Sandusky, to the village of Adrian in Lenawee county, Michigan, where it joins the Southern Rail-road. The proposed extension of this work diverges from the line at Palmyra, six miles from Adrian, and proceeds northward, through Tecumseh, Clinton, Manchester, Napoleon, and unites with the Central Rail-road at Michigan Centre. From the latter point to the town of Kalamazoo, the Central Rail-road will be used. At Kalamazoo the line is resumed and conducted along the valley of the Kalamazoo to the town of Allegan, where the location terminates. Length of the first section 75 miles—second section 28 miles, total length of the company's lines 103 miles; and entire distance from Toledo to Kalamazoo 183 miles.

YPSILANTI AND TECUMSEH RAIL-ROAD, 25 miles in length, leaves the Central Rail-road at Ypsilanti, and proceeding through Salem, intersects the Erie and Kalamazoo Rail-road at Tecumseh, and thus opens a rail-road communication, though rather circuitous, between Detroit and Toledo.

DETROIT AND PONTIAC RAIL-ROAD, extends from Detroit, the capital of the state, to Pontiac, the seat of justice of Oakland county; distant 25 miles north-west from Detroit.

ALLEGAN AND MARSHALL RAIL-ROAD, extends from Allegan to Marshall, passing through the villages of Bronson and Battle Creek. It lies in the counties of Allegan, Kalamazoo and Calhoun; is about 52 miles in length, and is now in course of construction by a joint stock company under the patronage of the state.

ST. CLAIR AND ROMEO RAIL-ROAD, from Palmer on the St. Clair to the village of Romeo in Macomb county. It was commenced in 1837 by a company chartered in 1836, with a capital of $100,000. Length 26 miles.

SHELBY AND BELLE RIVER RAIL-ROAD. Company incorporated in 1836 with a capital of $100,000. The road extends from Belle river, via Romeo, to Utica, 27 miles.

SHELBY AND DETROIT RAIL-ROAD, now in progress, extends from Detroit, the capital of the state, to Utica, via Shelby, in Macomb county. Length about 23 miles.

PALMYRA AND JACKSONBURG RAIL-ROAD, extends through Tecumseh, Clinton, Manchester and Sandstone, to Jackson, 46 miles. A great portion of the work is finished and the remainder in progress, under the patronage of the state.

RIVER RAISIN AND LAKE ERIE RAIL-ROAD, commences at Plaisance Bay, a few miles south-east from Monroe, through which it passes, and thence along the right bank of the Raisin to Blissfield, where it unites with the Erie and Kalamazoo Railroad. Length, including a branch from Dundee to Clinton, 50 miles.

AUBURN AND LAPEER RAIL-ROAD, 30 miles in length, from Lapeer in Lapeer county to Auburn in Oakland county, 30 miles long.

MOTTVILLE AND WHITE PIGEON RAIL-ROAD, extends from Mottville in St. Joseph county to the Indiana boundary.

MEDINA AND CANANDAIGUA RAIL-ROAD, extends from the town of Morenci, via Canandaigua and Medina, to the Southern Rail-road in Lenawee county.

Rail-roads are proposed from Detroit to Owasso : from Detroit to Utica, 23 miles : from Detroit to Monroe : from Monroe to some point on the Central Rail-road, and from Gibraltar to Clinton : from Romeo to Mount Clemens : from Detroit to Maumee Bay, via Monroe : from Ypsilanti to the River Raisin and Erie Rail-road : from Kalamazoo Village to Lake Michigan at the outlet of the South Black river, 25 miles in length : from Ann Arbor to Monroe, 33 miles : from Constantine to Niles, 33 miles : from Detroit to Shrawassee, via Farmville, Kensington, Howell and Byron, length 87 miles : from Saginaw to the Northern Rail-road, 40 miles long : from Gibraltar to Clinton, via Lisbon, 41 miles : from the head of ship navigation on the river Raisin to that of Grand river, below Grand Rapids, via Monroe, Tecumseh, Clinton and Marshall, length about 150 miles : from Mount Clemens, via Lapeer, to Saginaw, 91 miles

MICHIGAN. 219

long: from Clinton to Adrian 15 miles: from Kalamazoo to Lake Michigan, and from Saginaw to Leroy.

CANALS.

CLINTON AND KALAMAZOO CANAL. This is one of the proposed state works, not yet commenced. This canal which is intended to unite the waters of Lakes Michigan and St. Clair, will pass through the counties of Macomb, Oakland, Livingston, Ingham, Eaton, Barry and Allegan: and the towns of Mount Clemens, Rochester, Pontiac, Howell, Hastings and Singapore, its point of termination. The course of the canal as located is generally west, through the lakes of Oakland county and the valleys of Big Fork of Grand river and Rabbit river of Kalamazoo. The summit level (42 miles in length) is 344.61 feet above Lake St. Clair, and 336.11 feet above Lake Michigan. Difference of level between Lake St. Clair and Michigan, 8.50 feet. Total lockage, 690.72 feet. Estimated cost $2,250,000. Length 216 miles.

GRAND RIVER AND SAGINAW CANAL, is to extend from the north-east bend of Maple river, a branch of Grand river, to the Beaver Dam branch of the Shiawassee river, and thence by river navigation to Saginaw. Length 14 miles. Cost of canal and the necessary river improvements for steam-boat navigation, $238,240. There is also proposed, a short canal around the Falls of St. Mary, in the strait between Lakes Superior and Huron. Length, as proposed, 4,500 feet; depth 10 feet; width, in rock cutting, 50 feet, and the remainder 100 feet at the top water line; three locks 100 by 32; lockage 18 feet. Estimated cost $112,544 80.

Canals are proposed from Homer to Union City, 20 miles; from the Kalamazoo to Dexter; and the navigation of the Shiawassee and Huron rivers is to be improved by chartered companies.

Aggregate length of rail-roads in Michigan, 131.00 miles.

WISCONSIN TERRITORY.

WITH the exception of some surveys, authorized by Congress, nothing has been done in this section of the United States, in the way of canals or rail-roads. Rail-roads are proposed from Milwaukee, on Lake Michigan, to the Mississippi river; from Lafontaine to Winnebago Lake; from Belmont to Dubuque; from Belmont to Dodgeville, via Mineral Point. And a canal from Milwaukee to Black river, for which purpose Congress granted a tract of land in 1839.

IOWA TERRITORY.

THERE is no work of internal improvement yet commenced in this territory. Several rail-roads are spoken of, but the unsettled state of the country, and the derangements in its monetary affairs, will, no doubt, cause a suspension of active operations.

CANADA.

CANALS.

RIDEAU CANAL. This important work unites the waters of Lake Ontario with those of the Ottawa river. It commences at Kingston, on Lake Ontario, pursues a north-eastern direction, through a chain of lakes, with most of which it becomes identified in its course, until it intersects Rideau river. Continuing its route along the banks, and sometimes in the bed of that river, it enters the Ottawa at Bytown, a short distance above the mouth of the former, in north lat. 45° 23'. This highly important work, the existence of which is scarcely known in the United States, is now in active operation. Length from Kingston to Bytown on the Ottawa, including the natural navigable courses, 129½ miles; 53 locks, each 33 feet wide and 134 long. Ascent from Kingston to the Summit pond by 19 locks, 165 feet; descent from the Summit pond to the Ottawa, by 34 locks, 290 feet; total lockage 355 feet. Depression of the Ottawa below Lake Ontario at Kingston, 125 feet; general course, north-north-east.

WELLAND CANAL, is designed to open a navigable communication between Lakes Erie and Ontario. It leaves the former at Sherbroke, near the mouth of Grand river, crosses the Wainfleet Marshes to Chippewa river, and passes along its valley about ten miles. On leaving the Chippewa, the canal assumes a northern direction, traverses a deep cut, of nearly two miles in extent, and of the mean depth of 45 feet, and after a further course of 8 or 10 miles, enters Lake Ontario at Port Dalhousie, about 9 miles west of Niagara village. This splendid work, equalled in depth by the Chesapeake and Delaware canal only, is now completed. It admits the passage of the largest vessels that navigate the lakes, the dimensions of the locks north of the mountain ridge being 22 feet wide, 100 long, and 8 feet deep; those on the south of the ridge, 45 feet in width and 120 in length. Length from Port Maitland to Port Dalhousie, 36 miles; 34 locks, all descending. Descent, 334 feet; general course, north-east.

LA CHINE CANAL, 9 miles in length, extends from the southern suburbs of Montreal to the outlet of Lake St. Louis. This canal, with some others now in progress, are intended to overcome obstructions in the navigation of the St. Lawrence.

RAIL-ROADS.

LA PRAIRIE AND ST. JOHNS RAIL-ROAD. This rail-road, which is about 16 miles in length, forms a part of the principal line of communication between the States and Canada. It unites Lake Champlain with the St. Lawrence by a more convenient route than was formerly pursued.

A rail-road from Quebec to St. Johns, in New Brunswick, has long engaged the attention of the Canadian government and people. The line as proposed, on leaving Quebec, passes down the right bank of the St. Lawrence to St. Nicholas river, thence along its valley and across the dividing ridge into that of Black river, a branch of the St. Johns, whose valley is traversed in its whole course to the city of St. Johns, a distance of about 200 miles. A large portion of this line, it will be perceived, is within the limits of the state of Maine, as defined by the treaty of 1783.

A

CONDENSED SUMMARY

OF THE

CANALS AND RAIL-ROADS

IN THE

UNITED STATES;

THEIR LENGTHS, AND TERMINATING POINTS.

Canals in Maine.

Name	From	To	Miles.
Cumberland and Oxford,	near Portland,	Long Pond,	50.50

Rail-roads in Maine.

Bangor and Orono,	Bangor,	Orono,	10.00

Canals in New Hampshire.

Bow Falls,	.	.	.	.	.	0.75
Hookset Falls,	.	.	.	.	.	0.13
Amoskeag Falls,	.	.	.	.	.	1.00
Union,	.	.	.	.	.	9.00
Sewall's Falls,	.	.	,	.	.	0.25

Rail-roads in New Hampshire.

Name	From	To	Miles.
Eastern,	Mas. Line,	Portsmouth,	15.47
Nashua and Lowell,			15.00

Canals in Vermont.

White River Falls,	0.50
Bellows Falls,	0.16
Waterquechey,	0.40

Canals in Massachusetts.

Middlesex,	Boston,	Chelmsford,	30.00
Pawtucket,	Lowell,		1.50
Blackstone,	Providence,	Worcester,	45.00
Hampshire and Hampden,	Coun. Line,	Northampton,	22.00
Montague Falls,			3.00
South Hadley Falls,			2.00

Rail-roads in Massachusetts.

Eastern,	Boston,	N. H. Line,	40.00
Boston and Lowell,	Boston,	Lowell,	26.50
Andover and Wilmington,	Br.	Haverhill,	17.75
Charlestown,	Charlestown,		1.00
Boston and Worcester,	Boston,	Worcester,	44.00
Millbury Branch,		Millbury,	8.00
Great Western,	Worcester,	W. Stockb'e,	116.06
Boston and Providence,	Boston,	Providence,	41.00
Dedham Branch,		Dedham,	2.00
Taunton Branch,	Mansfield,	Taunton,	11.00
Taunton and New Bedford,	Taunton,	New Bedford,	24,00
New Bedford and Fall River,	N. Bedford,	Fall River,	13.00

SUMMARY.

Name.	From	To	Miles.
Sekonk,	Sekonk,	Providence,	5.00
Quincy,	Granite Q.,	Q. Landing,	3.00

Rail-roads in Rhode Island.

Providence and Stonington,	Providence,	Stonington,	47.00

Canals in Connecticut.

Farmington,	New Haven,	Mass. Line,	56.00
Enfield Falls,			5.50

Rail-roads in Connecticut.

Norwich and Worcester,	Norwich,	Worcester,	58.50
New Haven and Hartford,	New Haven,	Hartford,	40.00
Housatonic,	Bridgeport,	Mass. Line,	73.00

Canals in New York.

Erie,	Albany,	Buffalo,	363.00
Champlain,	West Troy,	Whitehall,	76.00
Chenango,	Utica,	Binghamton,	97.00
Black River,	Rome,	Carthage,	85.00
Oswego,	Syracuse,	Oswego,	38.00
Cayuga and Seneca,	Seneca lake,	Cayuga lake,	23.00
Crooked Lake,	Pennyan,	Seneca lake,	7.75
Chemung,	Seneca lake,	Elmira,	23.00
Branch, of do.	Elmira,	Knoxville,	16.00
Delaware and Hudson,	Eddyville,	Lackawaxen,	83.00
Genesee Valley,	Rochester,	Olean,	119.63
Dansville Branch,	Mt. Morris,	Dansville,	11.00

226 SUMMARY.

Name.	From	To	Miles.
Harlem,	. . .	Hudson river, East river,	3.00
Croton Aqueduct,	.	Croton river, N. York,	40.56

Rail-roads in New York.

Name		From	To	Miles
Long Island,	. .	Brooklyn,	Hicksville,	27.00
Harlem,	. . .	New York,	Harlem,	8.00
Hudson and Berkshire,	.	Hudson,	W. Stockbridge,	33.00
Catskill and Canajoharie,		Catskill,	Canajoharie,	78.00
Rensselaer and Saratoga,		Troy,	Balston,	23.50
Mohawk and Hudson,	.	Albany,	Schenectady,	15.86
Saratoga and Schenectady,		Schenectady,	Saratoga,	21.50
Utica and Schenectady,	.	Schenectady,	Utica,	77.00
Syracuse and Utica,	.	Utica,	Syracuse,	53.00
Syracuse and Auburn,	.	Syracuse,	Auburn,	26.00
Auburn and Rochester,	.	Auburn,	Rochester,	80.00
Tonawanda	. .	Rochester,	Attica,	45.00
Buffalo and Niagara Falls,		Buffalo,	N. Falls,	23.00
Lockport & Niagara Falls,		Lockport,	N. Falls,	20.00
Buffalo and Black Rock,	.	Buffalo,	B. Rock,	3.00
Rochester,	. .	Rochester,	Port Genesee,	3.00
Ithaca and Oswego,	.	Ithaca,	Owego,	29.00
Bath,	. . .	Bath,	Crooked lake,	5.00
Port Kent and Keesville,		P. Kent,	Keesville,	4.50

Canals in New Jersey.

Name	From	To	Miles
Delaware and Raritan,	.	Bordentown,	N. Brunswick, 42.00
Morris, . . .		Jersey city,	N. Easton, Pa. 101.75
Salem, ...		Salem creek,	Delaware river, 4.00

Rail-roads in New Jersey.

Name	From	To	Miles
Camden and Amboy,	.	Camden,	S. Amboy,	61.00
Trenton Branch,		. . .	Trenton,	8.00
Jobstown Branch,	-	Jobstown,	Craft's creek,	13.00

SUMMARY.   227

Name.	From	To	Miles.
Paterson and Hudson,	Jersey city,	Paterson,	16.30
Camden and Woodbury,	Camden,	Woodbury,	9.00
New Jersey,	Jersey city,	N. Brunswick,	34.00
Trenton and Brunswick,	Trenton,	N. Brunswick,	27.00
Morris and Essex,	Newark,	Morristown,	22.00
Elizabethport & Somerville,	Elizabethport,	Somerville,	25.00

Canals in Pennsylvania.

Pennsylvania Canal. {	Central Division,	Columbia,	Hollidaysburg,	172.00
	Western Division,	Johnstown,	Pittsburg,	104.25
	Susquehanna do.	Duncan's Is.	Northumberland,	39.00
	West Branch do.	North'land,	Farrandsville,	73.00
	North Branch do.	do.	Lackawana,	72.50
	Delaware Division,	Bristol,	Easton,	59.75
	Beaver Division,	Beaver,	Shenango R.	30.75
Schuylkill Navigation,		Philadelphia,	Port Carbon,	108.00
Union,		Reading,	Middletown,	82.08
Lehigh,		Easton,	Stoddartsville,	84.48
Lackawaxen,		Delaware R.	Honesdale,	25.00
Conestoga,		Lancaster,	Safe Harbor,	18.00
Codorus,		York,	Susquehanna R.	11.00
Bald Eagle,		West Br. Ca.	Bellefonte,	25.00
Susquehanna,		Wrightsville,	Havre de Grace,	45.00
Minor Canals				24.00

Rail-roads in Pennsylvania.

Columbia and Philad.,	Philadelphia,	Columbia,	81.60
Portage,	Hollidaysburg,	Johnston,	36.69
Philadelphia City, &c.			6.00
Valley,	Norristown,	Columbia R. R.	20.25
West Chester,	Columbia R. R.	West Chester,	10.00
Harrisburg & Lancaster,	Harrisburg,	Lancaster,	35.50
Cumberland Valley,	Harrisburg,	Chambersburg,	50.00
Franklin,	Chambersburg,	Williamsport,	30.00
York & Wrightsville,	York,	Wrightsville,	13.00

Name.	From	To	Miles.
Strasburg,	C. Val. R. R.	Strasburg,	7.00
Philad. and Reading,	Philadelphia,	Pottsville,	95.00
Little Schuylkill,	Port Clinton,	Tamaqua,	23.00
Danville and Pottsville,	Pottsville,	Sunbury,	44.54
Lit. Sch. and Susq.	Tamaqua,	Williamsport,	106.00
Beaver Meadow Br.	Lindner's Gp.	Beaver M. R. R.	12.00
Williamsp't & Elmira,	Williamsport,	Elmira	73.50
Corning & Blossburg	Blossburg	Corning,	40.00
Mount Carbon	Mt. Carbon,	Norwegian Cr.	7.24
Schuylkill Valley,	Port Carbon,	Tuscarora,	10.00
Branches of do.			15.00
Schuylkill,	Schuylkill,	Valley,	13.00
Mill Creek,	Port Carbon,	Coal Mine,	9.00
Mine H. & Sch. Haven,	Sch. Haven,	Mine Hill Gap,	20.00
Mauch Chunk,	Mauch Chunk,	Coal M.	9.00
Branches of do.			16.00
Room Run,	Mauch Chunk,	Coal M.	5.26
Beaver Meadow	Parryville,	Coal M.	20.00
Hazelton and Lehigh,	Hazelton M.	Beaver M. R. R.	8.00
Nesquehoning,	Nes'honing M.	Lehigh R.	5.00
Lehigh and Susq.	White Haven,	Wilkesbarre,	19.58
Carbondale & Honesd'le,	Carbondale,	Honesdale,	17.67
Lykens Valley,	Broad Mount.	Millersburg,	16.50
Pine Grove	Pine Grove,	Coal M.	4.00
Philad. and Trenton,	Philadelphia,	Morrisville,	26.25
Philad. Ger. & Norris.	Philadelphia,	Norristown,	17.00
Germantown Br.			4.00
Philad. and Wilmington,	Philadelphia,	Wilmington,	27.00

Rail-roads in Delaware.

N. Castle & Frenchtown, N. Castle, Frencht., Md. 19.19

Rail-roads from Newcastle to Wilmington and from Wilmington to Nanticoke Creek, are proposed.

Canals in Delaware.

Chesapeake & Delaware, Delaware City, Back Creek, 13.63

SUMMARY.

Rail-roads in Maryland.

Name.	From	To	Miles.
Baltimore and Ohio,	Baltimore,	Harper's Ferry,	80.50
Washington Branch,	Patapsco river,	Washington,	30.35
Balt. and Port Deposite,	Baltimore,	Havre de Grace,	36.00
Balt. and Susquehanna,	Baltimore,	York, Pa.	56.00
Reistertown Branch,	6 m. from Bal.,	Reistertown,	8.00
Wil. and Susquehanna,	Hav. de Grace,	Wilmingt., Del.,	32.00
Annapolis and Elkridge,	Wash. Br.,	Annapolis,	19.75

Canal in Maryland.

Chesapeake and Ohio,	Georgetown,	Hancock,	136.00

Rail-roads in Virginia.

Richmond, Fredericksburg and Potomac,	Richmond,	Aquia Creek,	75.00
Louisa Branch,	24 m. fr. Rich.,	Gordonsville,	49.00
Rich'd and Petersb'g,	Richmond,	Petersburg,	23.00
Petersb'g and Roanoke,	Petersburg,	Weldon,	59.00
Greensville,	Near Hicksf'd,	Gaston, N. C.,	18.00
City Point,	Petersburg,	City Point,	12.00
Chesterfield,	Coal Mines,	Richmond,	13.50
Portsm'th and Roanoke,	Portsmouth,	Weldon, N. C.	80.00
Winch. and Potomac,	Harp. Ferry,	Winchester,	32.00

Canals in Virginia.

Alexandria Canal,	Georgetown,	Alexandria,	7.25
James River and Kanawha,	Richmond,	Buchannan,	175.00
Dismal Swamp,	Deep Creek,	Joyce's Creek,	23.00
Branches,	.	.	11.00

Rail-roads in North Carolina.

Name.	From	To	Miles.
Wilmingt. and Raleigh,	Wilmington,	Weldon,	161.00
Raleigh and Gaston,	Raleigh,	Gaston,	85.00

Canals in North Carolina.

Weldon Canal,	Weldon,	Hd. Roanoke F's,	12.00
Club Foot and Harlow,	Club Foot cr.,	Harlow cr.,	1.50

Rail-roads in South Carolina.

South Carolina,	Charleston,	Hamburg,	135.75
Columbia Branch,	Branchville,	Columbia,	66.00

Canals in South Carolina.

Santee,	Cooper river,	Santee river,	22.00
Winyaw,	Kinlock Cr.,	Winyaw Bay,	7.40
Saluda,	Shoals,	Granby,	6.20
Drehr's,	Saluda Falls,	Head of Falls,	1.33
Lorick,	Broad River,	Head of Falls,	1.00
Lockharts,	Head of Falls in Broad Riv.	To Foot,	2.75
Wataree,	Jones's Mill,	Elliot's,	4.00
Catawba,	At var. points on the Catawba,		7.77

Rail-roads in Georgia.

Georgia,	Augusta,	De Kalb Co.	165.00
Athens Branch,	Georgia R. R.	Athens	33.00
Western and Atlantic,	De Kalb Co.	Tennessee R.	130.00
Central,	Savannah,	Macon,	193.00
Monroe,	Macon,	Forsyth,	25.00
Macon and Talbotton,	Macon,	Talbotton,	70.00

SUMMARY.

Canals in Georgia.

Name.	From	To	Miles.
Savannah, Ogeechee, and Alatamaha,	Savannah,	Alatamaha R.	16.00
Brunswick,	Alatamaha,	Brunswick,	12.00

Rail-road in Florida.

Wimico and St. Joseph,	Lake Wimico,	St. Joseph,	12.00

Rail-roads in Alabama.

Alabama, Florida, and Georgia,	Pensacola,	Montgomery,	156.46
Montg. and W. Point,	Montgomery,	West Point,	87.00
Tuscumbia, Courtland, and Decatur,	Tuscumbia,	Decatur,	44.00
Selma and Cahawba,	Selma,	Cahawba,	10.00
Wetumpka,	Wetumpka,		10.00

Canals in Alabama.

Muscle Shoals Canal,	Head of Falls,	Florence,	35.75
Huntsville,	Triana,	Huntsville,	16.00

Rail-roads in Mississippi.

West Feliciana,	St. Francisv.,	Woodville, (Miss. p.)	7.75
Vicksburg and Clinton,	Vicksburg,	Clinton,	54.00
Grand Gulf,	Grand Gulf,	Port Gibson,	7.25
Jackson and Brandon,	Jackson,	Brandon,	14.00

Rail-roads in Louisiana.

Name.	From	To	Miles.
Pontchartrain,	New Orleans,	Lake Pontch.	4.50
West Feliciana,	St. Francisville,	Woodv. (La. p.)	20.00
Atchafalaya,	Pt. Coupee,	Opelousas,	30.00
Alexandria, and Cheneyville,	Alexandria,	Cheneyville,	30.00
N. Orl. and Carrolton,	N. Orleans,	Lafayette,	11.25
Orleans Street,	N. Orleans,	B. St. Johns,	1.50

Canals in Louisiana.

Orleans Bank,	N. Orleans,	Lake Ponch.	4.25
Canal Carondelet,	N. Orleans,	B. St. John,	2.00
Barataria,	Near N. Orl'ns,	Berwick's Bay,	85.00
Lake Veret,	Lake Veret,	La Fourche riv.	8.00

Rail-roads in Tennessee.

La Grange and Memp.	La Grange,	Memphis,	50.00
Somerville Branch,	Moscow,	Somerville,	16.00
Highwassee,	Knoxville,	West and Atlantic R. R.	98.50

Rail-roads in Kentucky.

Lexington and Ohio R. R.,	Louisville,	Lexington,	92.75
Portage,	Bowlinggreen,	Barren river,	1.50

Rail-roads in Illinois.

Meredosia and Jacksonville,	Meredosia,	Jacksonville,	20.00
Coal Mine Bluffs,	Illinois,	Coal mine,	6.00

SUMMARY. 233

Canals in Illinois.

Name.	From	To	Miles.
Illinois and Michigan,	Chicago,	near Peru,	105.90

Canals in Indiana.

Wabash and Erie,	Lafayette,	Lake Erie,	187.00
Whitewater,	Lawrenceburg,	Brookville,	30.00

Rail-roads in Indiana.

Madison and Indianapolis,	Madison,	Indianapolis,	95.00

Canals in Ohio.

Ohio and Erie,	Portsmouth,	Cleveland,	307.00
Columbus Branch,	Columbus,	Canal,	10.00
Lancaster Branch,	Lancaster,	Canal,	9.00
Hocking,	Lancaster,	Athens,	50.00
Zanesville Branch,	Zanesville,	Canal,	14.00
Walhonding Branch,	Walhonding R.,	Canal,	23.00
Miami,	Cincinnati,	Defiance,	178.00
Warren Branch,	Middletown,	Lebanon,	20.00
Sandy and Beaver,	Bolivar,	Ohio river,	76.00
Mahoning,	Akron,	Beaver river,	77.00

Rail-roads in Ohio.

Mad R., & Sandusky city,	Tiffin,	Sandusky city,	36.00
Ohio,	Manhattan,	Sandusky city,	40.00

Rail-roads in Michigan.

Name	From	To	Miles.
Central,	Detroit,	Ann Arbor,	44.00
Erie and Kalamazoo,	Toledo,	Adrian,	33.00
Ypsilanti and Tecumseh,	Ypsilanti,	Tecumseh,	25.00
Detroit and Pontiac,	Detroit,	Pontiac,	25.00

GLOSSARY

OF THE

SCIENTIFIC, MECHANICAL,

AND

OTHER TERMS,

EMPLOYED IN ENGINEERING.

A.

Abrasion, rubbing off; the matter worn off by the attrition of bodies; a superficial excoriation of any part of a body.

Abutment, the solid mass of masonry at the ends of a bridge.

Acclivity, the slope or steepness of a line or plane, inclined to the horizon.

Adhesion, the force acting on the surface of two bodies in contact with each other, which prevents one sliding over the other.

Adit, a passage or entrance. The adits of mines are apertures by which they are entered, or the ores and water carried away.

Aggregate, the sum of several things added.

Alluvion, the gradual increase of land, not permanently submerged, along the sea shore, or on the banks of rivers, produced by the action of water.

Altitude, elevation, height in a vertical direction above a given base.

Angle of Repose, the utmost inclination at which a carriage will stand at rest upon a railway or road, and when, upon the least increase of slope, it is put in motion by the gravity of its weight. 1 in 250 or about 21 feet per mile, is considered the angle of repose upon a well constructed rail-road.

Apex or Vertex, the tip or point; the summit of any thing.

Aqueduct, a leader of water, built of stone or timber, to preserve its level, and to convey it from one place to another. It is applied either to a bridge over a valley, a road, or to a tunnel, when intended for the passage of water.

Arch, a portion of the circumference of a circle; a circular arrangement of overlapping stones or bricks, with radiating beds, commencing from fixed points or abutments, and meeting in the centre. Arches are of various shapes, semi-circular, segmental, eliptical or pointed.

Area, any plane surface, or the superficial contents of any figure.

Argillaceous, clayey, composed principally of clay.

Assistant engines, locomotives which are kept in reserve for assisting engines in ascending inclined planes.

Axis, a line about which a body may turn; the pin on which wheels revolve.

Axle, as applied to rail-road carriages, the transverse bar connecting the centres of the opposite wheels, with which it revolves, and to which it is fixed.

B.

Back-filling, the earth which is returned to its place, after the structure for which it was removed, is completed.

Backing, the top and side filling which sustains an arch.

Balance, a lever so adjusted as to determine the difference or equality of weights of bodies.

Ballasting, a term applied to the covering of roads, and to the filling in material above, below and between the stone blocks and sleepers upon rail-roads. It is mostly composed of gravel, broken stone, &c., and is laid about 2 feet thick on rail-roads. When the sub-soil of a rail-road is bad, a longitudinal drain 6 or 8 inches square is laid beneath the ballasting, with cross drains 12 or 15 inches apart, to convey the water into the side ditches, but stone ballasting seldom requires those drains.

GLOSSARY. 237

Barometer, an instrument for measuring the elasticity of air.

Basaltes, a black or greenish stone, consisting of prismatic crystals. It is always found standing in upright columns.

Base, the lowest part of any thing. In surveying, is a line on which a series of triangles is constructed, in order to determine the position of objects and places.

Basin, a recepticle for water, a region drained by a river.

Batter, the sloping face of a retaining, or other wall. The batter of a wall is either curved or straight; the average rate of the batter of the retaining wall on rail-roads, is $2\frac{1}{2}$ inches to the foot, and 1 inch for the wing walls of bridges.

Bearings, the chairs which support the frame work which rests upon the axles of a rail-road carriage.

Bench, a ledge left on the face of a cutting to strengthen it; benches are usually made at a change of slope, occasioned by meeting with a change of soil.

Bench marks, in surveying, are fixed points left along the line of survey for future reference.

Berm, is that bank or side of a canal which is opposite to the tow path.

Beton, a concretion used in foundations of hydraulic works. It consists of 12 parts of puzzolana, 9 of quick-lime, 6 of sand, 13 of stone scrapings, and 3 of iron scales from the smith's forge.

Bitumen, mineral pitch.

Bituminous shale, an argillaceous shale or indurated slaty clay, highly charged with bitumen.

Bisection, the division of a quantity into two equal parts.

Block, (stone) a support or foundation for the tracks or rails of rail-roads, upon which the chairs are secured.

Bond, the union and tie of the several stones or bricks in a wall.

Boring, a vertical sinking made in the earth by an augur, for the purpose of obtaining water.

Brace, an iron plate employed to strengthen the joinings of wood and other work.

Break, a lever attached to a locomotive engine or rail-road carriage, which presses upon the rim of the wheels and regulates the velocity.

GLOSSARY.

Breakwater, a mole or projection, designed for the protection of shipping against the violence of storms.

Brick, a preparation of clay, sand and ashes, burnt in a kiln. Good brick earth is often found in a natural state. A brick is 9 inches long, $4\frac{1}{2}$ wide and $2\frac{1}{4}$ thick.

Bridge, a common engineering contrivance for passing over rivers, canals, &c.

Buffing apparatus, an expedient for receiving the shock of a collision between rail-road carriages, consisting of powerful springs and framing.

Buffer-head, the box fixed at each end of the rods connected with the buffing apparatus, which receives the shock, and communicates it to the springs.

C.

Caisson, a large water tight box used for the purpose of placing the foundation of piers, &c. It is sunk to the bottom of the water and the masonry commenced within it and carried up to the surface of the water, when the sides are removed, leaving the pier resting on the bottom of the box.

Calcarious Rock. Limestone.

Canals, an artificial cavity in the earth, filled with water to afford an easy and cheap conveyance for goods, &c. Canals for the transportation of merchandize, are constructed of various capacities, according to the extent and nature of the trade they are intended to accommodate. Those of Great Britain and the United States are generally four feet deep and sixty feet wide at the top water line; but in both countries, there are canals which differ materially from the usual dimensions. The Caledonian Canal of Scotland, for example, which is designed for the passage of ships of war, is twenty feet deep and one hundred and twenty-five feet wide; whilst others are not more than ten feet in width and of a corresponding depth. The mode of constructing canals is nearly the same every where; but as most persons are unacquainted with the leading principles on which they are formed: we shall offer a few remarks on this head. Previously to the commencement of a canal, the engineer decides upon its dimensions, which, of course, depends upon the nature and probable extent of the trade for which it is designed. In settling and arranging the form and principles of construction, many nice calculations, requiring powers of a high order, are necessary. These relate, chiefly to inquiries, relative, 1st. to the requisite supply of water; 2d, to the extent of excavation and embankment and cost of construction generally; and 3d, to the form and size of boats best adapted to overcome the resistance which all canals, in a greater or lesser degree, offer to the passage of boats. With regard to the first, it is usual to ascertain the probable supply that can be obtained under the least favourable circumstances. To do this, it is necessary to find the mean temperature of the region through which the canal is to pass; the perpendicular inches of water which is equivalent to the mean quantity of aqueous vapour suspended in the air at that temperature, and the mean annual depth of rain in perpendicular inches. In the latitude of Philadelphia, the mean temperature is about 62°: aqueous vapour 3,968, and the annual rain 34.92 inches. The minimum or least quantity of water which runs in rivers varies much more than the mean quantity. If the river is very short, and drains a basin composed of gravel and sand, it will not be subject to floods, and the least discharge may be proportionally equal to one-third of the whole quantity of rain which falls in its basin. If the river is very long, passing from a cold to a warm climate, having a basin generally formed of clay, and having no lakes in its course to regulate its discharge, it will be subject to high floods, and the minimum discharge may be no more than $\frac{1}{100}$ part of the mean rain. The basin of the Mississippi, including the Missouri, contains about 1,210,000 square miles; the mean annual quantity of rain is about 101,198,592,000,000 cubic feet; the mean annual discharge is about 30,000,000,000,000 cubic feet, or one-third of the rain; and the mean minimum discharge is about 20,000,000,000,000 cubic feet, or one-fifth of the rain. The basin of the Po, in Italy, contains 45,600 square miles; the mean annual rain is 3,813,765,120,000 cubic feet; the mean annual discharge is 1,850,112,000,000 cubic feet, or about one-half of the rain. The basin of the Tay, in Scotland, contains 2,315 square miles; the annual rain is about 130,000,000,000 cubic feet; the annual discharge is 100,000,000,000 cubic feet, or four-fifths of the rain. At a very dry time it was reduced to 457 feet per second, or one-ninth of the rain. For the purpose of filling reservoirs, the mean annual discharge of rivers, subject to some deduction, is relied on. For the purpose of filling canals from the natural flow of the stream, the minimum discharge only is taken into the calculation. The quantity of water running in large rivers, is commonly estimated in the following manner. The area of a transverse vertical section of the stream being determined, the mean velocity of its current is ascertained by observing the time in which a body moves through a given space in various parts of the stream, on and below the surface. The average time compared with the distance, gives the apparent velocity. If the stream is shallow, not more than fifteen inches deep, with an uneven bottom, the mean velocity will be about eight-tenths of its apparent velocity. If, on the contrary, the river is four or five feet deep, with a smooth bottom, the mean velocity will be nearly equal to its apparent velocity. From these data the discharge in any given time may be deduced. For small streams the process is more simple. It consists in causing all the water to pass through an aperture, the dimensions of which being known and the velocity ascertained, the amount of water discharged, is readily computed.

On the subject of excavation, embankment, &c., the quantities are calculated by the rules of mensuration and set down in cubic yards. Tables are prepared, by which the quantities in each chain are found by inspection. The cost of a canal depends greatly upon the nature of the soil and climate; the price of labour and provisions and the other local circumstances. That of excavation increases with the depth, but not in direct proportion. For a prism three feet in depth through sand or gravel, the expense is about six cents per cubic yard, and for clay or stony matter, eleven cents. Excavation of loose rock costs about fifty cents, and of solid rock one dollar per cubic yard. These prices are very low and can only be assumed under the most favourable circumstances. The cost of embankment generally exceeds that of excavation. The problem of the relative amount of resistance opposed to the passage of boats by canals of various dimensions, is one, though deserving close attention, not always duly considered in deciding upon the capacity of canals. It has been ascertained that the amount of resistance to boats thirteen feet and a half wide and three feet draft, in a canal sixty feet wide and six deep, is 81¼ pounds; in a canal forty-eight feet wide and five deep, 100 pounds, and in one forty feet wide and four deep, it is 130 pounds; and that 18 per cent. more weight can be transported by the same power on the former than on the latter canal.

The canals in the United States are generally four feet deep, with the inner slopes 1¾ to 1, and outer slopes 1½ to 1; towpath bank ten feet wide at top, and the berm six feet. If the ground is level, 2.75 feet of excavation is made, which, when thrown upon the banks, elevate them so as to form the requisite depth (four feet) for the prism. When ground is not perfectly level, three feet of cutting is assumed as the standard. As it may be necessary to cut deep or to prolong the line in order to maintain the grade, and as the ratio of the quantities in proportion to the depth, as well as that of prices, is constantly increasing, it is obvious that a proper medium between such extremes should be aimed at.

The walls of locks and the arches of culverts and aqueducts are usually built of stone. As the business on canals, especially those of our own country, is likely to be permanent, it is of infinite importance that they be finished in such a manner as to prevent the evil consequences resulting from defects in their construction. Much depends upon the original location of a canal, as well as upon the selection and combination of the materials. It may be made so *expensive* as to be unproductive; it may be made *cheap*, but so long and circuitous, as to discourage trade, and it may be so imperfectly made as to require its entire income to keep it in a condition for use. Cases, such as these, are by no means uncommon in this country: and hence the failure of many works of this description, which, under judicious and economical management, might have realized all the anticipations of the proprietors.

Car, a machine for the conveyance of passengers and goods on rail-roads. They are mounted on wooden frames above the wheels, the bearings of the axles being on the outside. They are protected from the effects of concussions by the buffing apparatus.

Carbon, an undecomposed inflammable substance.

Catenarian curve, a mechanical curve which a chain or rope forms itself into by its own weight, when hung freely between two points of suspension whether those points be in the same horizontal plane or not.

Cement, a composition of certain mineral substances, capable of uniting and keeping things together in close cohesion.

Centre of gravity, is that point in a body, which body, if supported, will be in equilibrio in every position.

Centre (of an arch), the wooden frame or mould used in the construction of arches in supporting the arch stones whilst in progress of construction.

Chair, a pedestal or socket of cast iron, used on rail-roads, for receiving and securing the rails, generally weighing from 12 to 20 lbs. each. Chairs are secured to the blocks by oak trenails and iron pins, a hole being first drilled in the blocks two inches in diameter, into which the trenails are driven; a $\frac{3}{4}$ inch hole is then bored into the latter, and the iron pin passed through the seat of the chair, and driven into the trenail. A piece of felt is introduced between the block and chair to ensure a firm bearing; the chairs are also fastened to the sleepers by pins.

Chord, is the right line which joins the extremities of an arc of a circle.

Circumferentor, an instrument used by surveyors in taking angles.

Clamps, thick planks used to sustain the ends of beams and prevent separation.

Clay-slate, a common mineral, chiefly composed of silica, alumina, peroxide of iron, and carbon.

Cleats, pieces of wood of different shapes, used to fasten ropes upon.

Coal Formation or Coal Field, a region in which coal abounds.

Coffer Dams, circular, oblong, or oval inclosures formed for the protection of workmen while engaged in laying the foundation of piers and other works in water. They consist of a double row of sheet piling, bolted together, enclosing a large body of clay well pemmed in, having stays, raking piles, and braces at the back of the same, to support the pressure of the water from the outside. When the dam is finished, the water is pumped out and the masonry commenced.

Cohesion, that kind of attraction, which, uniting particle to particle, retains together the component parts of the same mass.

Coke, a preparation of fossil coal, which is deprived of the naptha, bitumen or asphaltum it may have contained.

Combined Locks, canal locks placed side by side so as to admit of the ascent and descent of boats at the same time.

Compass, an instrument employed by surveyors and others to ascertain their courses.

Composite Locks, canal locks built with stone faced with timber.

Concrete, a mixture composed of cement, broken stone and gravel, or lime, gravel and sand, 1-7 to 1-9 of lime.

Cone, a solid body having a circular base, and its other extremity terminated in a single point or vertex.

Conduit, a canal of pipes for the conveyance of water or other liquids.

Configuration, the exterior surface or shape that bounds bodies and gives them their particular form.

Conglomerate, rounded waterworn fragments of rock or pebbles cemented together.

Coping, the upper tire of masonry which covers the wall.

Counterfort, the pier or buttress which supports a retaining wall, sometimes used as a tie to the material at the back of the wall.

Countersink, one cavity corresponding in size and position with another.

Construction, the act of building.

Continuous bearings, track-ways constructed of timber beams or sleepers placed longitudinally of the road-way, and pinned upon transums. The system of continuous bearings is much used in this country, on account of the abundance of timber.

Convex, round or curved or protuberant outwards, as the outside of a globular body.

Cramp, an iron tie used for securing the stones of a wall together. A vertical cramp is termed a dowel or plug.

Crane, a machine used for raising and lowering heavy weights.

Crank, a square piece projecting from a spindle, serving by its rotation to raise and fall the pistons of engines.

Crossing, a communication from one rail-track to another. They are similar to sidings, having switches and crossing-points.

Cross-sills, blocks of stone or wood laid in the broken stone filling, which support the sleepers.

Cross-ties or Sleepers, are pieces of timber laid at intervals of two or three feet, across the road-way, upon which the string pieces and rails are placed. They are intended to keep the rails in their proper position. In some cases the chairs which support the iron rails are placed upon the sleepers.

Crow, a kind of iron lever, sharp at one end, used for heaving or pushing great weights.

Culminate, to be vertical, to be in the meridian.

Culvert, a passage or arch-way for water, under a canal or rail-road.

Curvature of a line, is its bending or flexure by which it becomes a curve of any particular form and proportion.

Curve, is a line, the several parts of which proceed in different directions. A bend in a line of rail-road.

Small curves on rail-roads should be avoided, as the centrefugal force acquired by the train has a tendency to throw it off the track.

Curvelinear, any thing relating to curves.

Cut, a notch in the cross-sleepers for the reception of the wooden rail.

Cutting, a term frequently applied to excavations.

D.

Dam, a mole or bank to confine or regulate the flow of water.

Damage, an injury or hindrance attending a person's estate.

Damages, (land,) compensation for injury done to lands, buildings, &c., in the construction of canals and rail-roads.

Damper, a contrivance to check and control the speed of a locomotive.

Data, denote certain quantities which are given or known,

and by means of which other quantities which are unknown, are to be determined.

Datum Line, is the base or horizontal line of a section, from which all heights and depths are calculated, which has reference to some fixed point in the line.

Decomposition, the separation of the constituent parts of bodies.

Defile, a narrow passage.

Deflection, the turning any thing aside from its former course.

Degree, the 360th part of the circumference of any circle.

Deep Cut, any open excavation of unusual depth.

Density, vicinity, closeness, compactness.

Depot or Station, as now understood, denotes the stopping place for rail-road cars, &c.; tool houses, &c.

Depression, a sinking below a common plane.

Descent, implies a downward motion.

Deviation, change of route. In England, deviations within 300 feet from the line as originally loacted, are allowed by special act of parliament.

Diagonal, a right line drawn across a quadrilateral or other figure, from the vertex of one angle to that of another.

Diagram, is a scheme for explanation or demonstration of any figure or of its properties.

Diameter, the line which divides a circle into two equal parts.

Ditch, a trough for the reception of water, a drain.

Divergent, tending to various parts from one point.

Draining Tiles, are sometimes used in embankments to divert and carry the water off to the side drains.

Draining, the process of clearing wet and boggy lands from their superfluous moisture.

Drain, a ditch to draw water from low grounds.

Drawbridge, a bridge made after the manner of a door, to draw up or let down, to allow vessels to pass from one side to the other.

Dredge, a machine for clearing out canals, deepening rivers, &c.

Drift, a passage dug under the earth between the shafts in mines.

Drum or Rope Roll, a cylinder, generally of cast iron, used on inclined planes, for the purpose of conveying carriages up or down the plane. Drums are used when the plane is worked by a single rope.

Dyke, a work raised to oppose the passage of water.

E.

Earthwork, is a term applied to cuttings, embankments, &c.

Edge Rail, is an iron bar or girder, upon which the periphery of the wheels revolve, a flange being formed upon the inner edge of the same, projecting one inch to prevent their sliding off the rail. This description of rail was originally made of cast iron of various lengths, with a flat base at each end, in which holes are left for the insertion of pins, by which it is secured to the sleeper. Cast iron chairs were ultimately adopted for this purpose. The rail was fish-bellied on the under side; which form they have retained until recently; the head being made about $2\frac{1}{2}$ inches wide, and rounded, and a cross section taken through the centre of a rail, exhibited a greater thickness of metal at the upper part than the lower. The wrought-iron edge rail was afterwards employed, consisting of merely flat bars at first, from one to two inches square; or bars one or two inches by three inches; which, owing to their rounded heads and narrow shape, damaged the wheels. These were used until a method of rolling and manufacturing iron rails of a fish-bellied form, with heads complete, similar to the cast iron rail, was adopted some years since.

Embankment, a mound of earth thrown up to maintain the grade of a canal, rail-road, &c.

Engine. See Steam Engine.

Engineer, the conductor of a steam engine.

Equable Motion, is that by which the moveable body proceeds with the same continued velocity.

Evaporation, is a conversion of water into vapour, which becoming lighter than the atmosphere, is carried far above the earth's surface.

Excavation, the act of cutting or digging into hollows; removal of earth.

Explosion, a sudden and violent expansion of an ærial or other elastic fluid.

Extension, prolongation or augmentation in the length of a line.

Extrados, the outside of an arch.

F.

Face, Façade, that superficies of a structure, which lies in front.

Feather Edge, wedge shape.

Feeder, side cuts which lead from streams or reservoirs, into, and supply, canals with water. A "navigable feeder" is one of sufficient capacity to admit of the passage of boats.

Felt, a fabric of hair and wool worked into a firm texture. It is sometimes used on rail-roads between the under side of the chairs and the upper surface of the blocks, to secure a firm hold.

Fencing, an enclosure for the protection of canals, rail-roads, &c.

Ferruginous, any substance partaking of the nature of iron or that contains particles of that metal.

Fissures, are crevices or clefts that divide the several strata of which the earth is composed.

Flange, the inner rim of a wheel, intended to confine it within certain limits.

Flood, the rising of waters.

Floor, platform, level area.

Fluid or Fluid Body, is that whose parts yield to the smallest pressure.

Force, strength, power. The agents which are usually employed as the first movers of machinery on a large scale, are the strength of men and animals; the force of falling water; the force of wind; and the force of steam or other elastic fluids.

Formation, the rocks and other solid arrangement of matter, of which the globe is composed.

Forebay, is that part of a mill-race through which the water flows upon the wheel.

Fossil, any thing dug out of the earth; organic remains.

Forcing Pump, a machine which raises water by alternate

motions. It has a solid piston, which, after the water has passed the lower valve, is forced down, and thus causes the fluid to pass into the conducting pipe, where there is also a valve to prevent its return. The first valve closes as the piston descends, while the second valve rises to allow its escape from the main pipe. When the piston rises the water follows through the first valve, while the other is closed by the superincumbent water in the conduit, and by the attraction of the piston, the water rushes after it to prevent a vacuum.

Foundation, is that part of a structure upon which all erections repose; it is usually under ground.

Free Stone. See Sand Stone.

Friction, is the retarding force produced by the cohesion of bodies, and the resistance of rough surfaces which are moved upon each other. It is in proportion to the pressure, and not to the surfaces in contact; but increases with the time in contact before they are moved. The friction of oak wood moving upon oak in the direction of the fibres, is rather more than one-third of the pressure, and across the fibre, it is rather less than one-third. With iron moving on iron, the friction is two-sevenths, and with iron on brass, it is one-fourth of the pressure. The friction when the surfaces are greased with tallow, is as follows:—iron upon brass it is one-eighth; iron upon oak, one twelfth; oak upon oak, one twenty-fifth. Friction is not sensibly influenced by different velocities.

The friction or resistance of the wheels of a carriage, arises first, from the *friction of attrition*, or the pressure of the axles against the bearings resting upon them, which support the carriage; and secondly, from the rolling friction, or the resistance offered to the revolution of the wheels, by the roadway; the amount of which principally depends upon the degree of smoothness and hardness of the surface over which the wheels are run. The friction of the axles forms by far the greatest resistance, and it is very important to keep up a constant supply of oil, to reduce it as much as possible.

Fuel, the matter or aliment of fire.

G.

Gage, (of a condenser,) is designed to exhibit the exact density and quantity of the air contained at any time in the condenser of a boiler.

Galena, sulphuret of lead.

Gallery, a term applied to underground excavations. Tunnels are sometimes worked by horizontal shafts, termed galleries, instead of vertical openings.

Gas, a permanent elastic aeriform fluid.

Gases, (noxious) the gases formed in tunnels by the fire of locomotives.

Geology, is the doctrine of the earth in its insentient or unorganized frame ; or of those masses of rock, strata, minerals, &c. of which it is composed.

Gneiss, a stratified primary rock, composed of the same materials as granite, but having usually a larger proportion of mica, with a laminous texture.

Graduation, the act of modifying or adjusting a roadway into a particular line. In rail-road making, it signifies the process by which a required grade is obtained.

Grade, the condition of a rail-road when prepared for the reception of the superstructure.

Grade line, or profile, is a prescribed line which governs the construction of a rail-road.

Gradient or *clivity*, denotes the proportionate ascent or descent of the several planes on a rail-road ; thus for example, an inclined plane four miles long, with a total fall of 36 feet, is described as having a descending gradient or clivity of 1 in $586\frac{2}{3}$, or a fall of 9 feet per mile.

Granite, one of the primary rocks, composed of quartz, felspar and mica.

Gravel, a congeries of small pebbles.

Gravity, is that mysterious but essential property of matter, by which every particle is attracted or drawn towards every other particle, inversely to the squares of their distances ; hence all bodies near the surface of the earth, have a tendency to move or fall towards its centre, and this tendency or force is called their weight.

Gravity, as applied to rail-roads, refers to the extra weight acquired by a train of carriages, when upon planes not perfectly level, or the force of the downward pressure, which is in proportion to the gradient of the plane. If the train is proceeding up the plane, great additional power is necessary to overcome the gravity, compared with that required upon a level, particularly if the same velocity is to be maintained. According to Brees, the resistance by gravity on a plane 1 in 50, is 44.80 lbs. per ton, and on 1 in 90 it is 24.88 lbs. per ton, which on a train of 60 tons gross, amounts to 1493 lbs., which is sufficient force to propel a train amounting to 186 tons upon a level.

Grit, a form of argillaceous earth, with a texture more or less porous, equable and rough to the touch.

Gross, the entire quantity, without deduction.

Grouting, a kind of liquid mortar, floated over the upper beds throughout a course of masonry or brick work.

Group, in geology, signifies an assemblage of various rocks, &c.

Grubbing, clearing the ground from trees, rocks and other impediments, preparatory to the commencement of a canal, rail-road, &c.

Guard lock, in canalling, is employed in maintaining the level of a canal, by preventing the encroachment of water from rivers, lakes, &c., when elevated beyond the prescribed level.

Gypsum, or sulphate of lime, a mineral composed of lime and sulphuric acid.

H.

Headway, the space under an arch or other structure.

Hornblende, in mineralogy, a species of clay. The common hornblende is of a greenish black, which in some varieties, approaches to a grey, and sometimes to a velvet black.

Horse power, the power of engines is estimated by comparison with the amount of force exerted by a horse, which is generally reckoned equal to 33,000 lbs., raised 1 foot high per minute, and if continued throughout the whole day of 8 hours, amounts to 150 lbs. conveyed a distance of 20 miles, at the rate of $2\frac{1}{2}$ miles an hour; but some engineers consider 125 lbs. a sufficient load for a horse.

H rail, a rail, which when cut transversely, presents the form of an H. See *Rail*.

I.

Inclination, denotes the mutual approach or tendency of two bodies or planes towards each other.

Inclined plane, is a plane surface, which forms with the horizontal plane any angle whatever; such angle is called the inclination of the plane. Inclined planes on rail-roads are designed to overcome inequalities in the profile; they perform the same office for rail-roads, that locks do for canals, ascend or descend from one level to another.

Indurated, hardened, compact, being hard.

Inflection, or deflection, the act of turning or bending.

Intermediate space, in double track rail-roads, means the space between the two inner lines, usually the same as the width of the track, 4 feet $8\frac{1}{2}$ inches.

Internal, in general signifies whatever is within a thing.

Intrados, the interior curve of an arch.

Invert, to turn any thing the contrary way.

J.

Joinings, in construction, those parts of a structure which are united and held together.

Joint chairs, a chair which secures the connection of two rails. It is usually larger than the common chair.

K.

Key or *Cotter,* a wedge-shaped or tapering piece of iron or wood, which is driven firmly into a mortice prepared to receive it, to tighten the several parts of the framing together, as a rail to a chair, &c. forming a fastening.

Kyanize, a process of preserving timber from dry rot, invented by Mr. Kyan. It consists in the use of a solution of corosive sublimate, in which the timber is immersed, and which neutralizes the primary elements of fermentation and renders the fibre of the wood indestructible. It also is said to season the timber, occupying only two or three months, instead of as many years by the old drying process. Dr. Earle of Philadelphia, has recently invented a method of preserving timber by a more

L.

Lamina, plates usually applied to the smaller layers of which a stratum is composed.

Lateral, projecting from the side.

Leakage, is the quantity which runs out of a defective cask or other vessel.

Ledge, a ridge of rocks; any prominence.

Level, even; not having one part higher than another; a horizontal line.

Level, an instrument employed in ascertaining the variations in the height of grounds, and in taking the section of a line of road or canal.

Level crossing, is when a rail-road crosses roads upon the same level.

Levelling, is the act of ascertaining a line parallel to the horizon at one or more stations, to determine the height or depth of one place with respect to another.

Lever, an inflexible bar which is capable of moving freely round a fixed point, called its fulcrum. It is used to overcome forces, and to elevate great weights, to which one end of the lever is applied, while the impelling power is applied to the other.

Levity, the opposite of gravity; or that quality in certain bodies, which gives them power to ascend.

Lift Locks, are those sections of a canal inclosed between two gates, which, on being filled with water or emptied, elevate or depress the boat, and thus allow it to pass from one level to another. When a boat is to pass from a higher to a lower level, it is floated into the lock and the gates closed; the water is then allowed to escape from the lock chamber to the lower level, which is effected either by paddles formed in the gates or by side culverts; the boat being thus sunk to the lower level, the lower gates are opened, when it passes through. Boats are passed up by the same process, reversed. The difference between the levels is termed the *lift of the lock*, which range

GLOSSARY.

from 3 to 30 feet. That portion of the lock enclosed between the gates, is called the *lock chamber*, which varies in capacity according to the trade which it is designed to accommodate.

Lime, calcined limestone, which is burned in kilns, after which process it is called *quick lime;* upon applying water it instantly expands and cracks, producing a considerable degree of heat; it then falls into a powder, when it is called *slacked lime.*

Limestone, the native indurated carbonate of lime.

Location, position; act of placing.

Lock, the general name for all those parts of a canal made to confine and raise the water.

Lockage, means the rise or fall effected by a lock or series of locks.

Locomotive engines, those engines which effect their own progressive motion by means of their internal machinery and the adhesion of their wheels to rail-road tracks.

Lode, a vein or course, whether metallic or not; a term most commonly used in mining operations.

Lode, in mineralogy, a metallic vein in rock formations.

M.

Machine, signifies any thing used to augment or regulate moving forces or powers; or it is any instrument employed to produce motion in order to save either time or force.

Main, a leader of water; pipe.

Marl, is a combination of alumine, silex and alum, and is denominated calcareous, argillaceous or silicious, as the lime, clay or silex, is most abundant.

Masonry, a term applied to all works of which stone is the chief material.

Matrix, or *Mother earth,* the stone in which metallic ores are found enveloped.

Maximum, denotes the greatest quantity attainable in a given case.

Mechanical power, force employed to overcome resistance; it comprehends steam, water, man and horse power.

Mensuration, that branch of mathematics which treats of

the measurement of the extensions, capacities, solidities, &c. of bodies.

Meridian, is a circle in the terrestrial sphere, which passes through the poles of the earth.

Meridian (First,) is that from which all the others are reckoned : Paris is the first meridian of France; Greenwich of England, &c.

Mica, sometimes but erroneously called *talc*. A simple mineral, having a shining silvery appearance, and capable of being split into very thin elastic leaves or scales.

Mine, a place under ground from whence metals, minerals, or even precious stones, are procured.

Minimum, the smallest amount.

Mineralogy, is that science which treats of the solid and inanimate materials of which our globe consists.

Mitre, a mode of joining two boards together.

Mole, a long pier or artificial bulwark of masonry, extending across the entrance of a harbour, in order to break the force of the sea.

Momemtum, in mechanics, is the same with impetus, or quantity of motion; and is generally estimated by the product of the velocity and mass of the body.

Mortar, a cement used in building, composed of lime and sharp coarse sand, and sometimes hair.

Mortice and Tenon, a joint used in wood. The extremity of one piece of timber is let into the face of another piece,—a tong being formed at the end of the piece to be let in, which is called *tenon*, and the hole cut in the face of the other is termed a *mortice*.

Motive power, the propelling force by which motion is obtained.

Mound, an artifical mount; an eminence.

N.

Navigation (inland or internal), expresses those means of intercommunication which are afforded by canals, rivers, lakes, &c.

O.

Oblique Arch, the heads of the courses of an oblique arch, consist of spiral lines, wound round a cylinder, every part of which cuts the axis at a different angle, the angle being greatest at the keystone and least at the springing; when they are wound round the cylinder, and viewed from beneath, they present the appearance of straight lines.

Offsets, in surveying, are those short perpendiculars that are measured on the sides of irregular figures, for the accurate determination of an area.

One in ten, when applied to inclined planes, means one foot vertical and ten horizontal.

One to one, a slope of 45° is said to be one to one, and so on.

Ores, metals, when found in a state of combination with other substances, are called ores.

Organic remains, (Oryctology), animal and vegetable substances which are dug out of the earth in a mineralized state.

Overfall, the brow of a dam, or that part over which the water passes.

Oxide, the combination of a metal with oxygen; rust is the oxide of iron.

P.

Palisades, stakes made of strong split wood used for the support of embankments.

Parallel motion, is a term denoting the rectilinear motion of a piston rod, &c. in the direction of its length; and contrivances by which such alternate rectilinear motions are converted into rotatory ones, and *vice versa*, in pumps, steam-engines, saw-mills, &c.

Parapet, an elevation or wall, breast high, on the sides of bridges, &c.

Pebbles, the name of a genus of fossils, distinguished from the flints by having a variety of colours.

Periphery, the circumference of a circle, ellipsis or any other curvilinear figure.

Perpendicular, is formed by one line meeting another, so as to make the angles on each side of it equal to each other.

Piers, in bridging, are the walls built to support the arches, and from which they spring as bases.

Pilaster, a square column, sometimes insulated but more frequently let within a wall, and only showing a fourth or fifth of its thickness.

Piles, are large stakes or beams sharpened at one end and shoed with iron, to be driven into the ground for a foundation to build upon in marshy places.

Pin or *Pinning in*, a system of wedging or underpinning the bed of a stone wall, when not properly squared, to supply any deficiencies.

Pipes, conduits for the conveyance of water.

Pipe chamber, a receptacle for water.

Piston, a cylinder of metal, fitted exactly to the cavity of the barrel or body of the pump; there are several kinds of pistons, some with valves, and others without; the latter are called *forcers*.

Pitman, an appendage to a forcing pump.

Pivot, a fort or shoe of iron, or other metal, usually conical, or terminating in a point, by which a body, intended to turn round, bears another fixed and at rest, and performs its revolutions.

Plane, a smooth surface, or one that lies evenly between its bounding lines. Applied to rail-roads, it refers to each length of a line of railway at the same gradient or inclination: they are of two kinds, level and inclined.

Planking, the act of covering and lining the sides and bottom of a canal, &c. with wood.

Plate rail, a flat bar of iron usually 15 feet in length, $2\frac{1}{2}$ inches wide, and $\frac{1}{2}$ or $\frac{2}{3}$ of an inch in thickness. This kind of rail is most commonly used in streets.

Platform, is a number of planks laid together, forming a kind of floor or terrace.

Plug, a piece of timber or metal formed like the frustrum of a cone.

Plunger, in mechanics, the same with the forcer of a pump.

Pool, a lake or pond formed in the bed of a stream by means of a dam.

GLOSSARY. 257

Prism, in geometry, is an oblong solid, contained under more than four planes, whose bases are equal, parallel and alike situated.

Profile, the outline of a figure, the top line of a rail-road or canal.

Protection wall, is usually employed to shield the banks of a canal, &c. from the effects of rains.

Puddle, a mixture of clay rendered impervious to water, and used for the purpose of excluding water from any works.

Puzzolana, a natural cement consisting of volcanic ashes, much used in hydraulic works; the addition of a small portion of lime hardens it quickly, even when laid under water.

Pyramid, in geometry, is a solid having any plane figure for its base and triangles for its sides, all terminating in one common point.

Q.

Quartz, silex or earth of flints.

Quoins, in architecture, denote the corners of brick or stone walls.

R.

Radius, the semi-diameter of a circle, or right line drawn from the centre to the circumference. In trigonometry, the radius is termed the whole sine or sine of 90 degrees.

Rail-road, is a track composed of wood, stone or iron, or a combination of all these materials, intended to diminish friction and for the more easy conveyance of heavy loads. Until within a few years rail-roads were mostly confined to mines of various descriptions, but they are now coming into general use. The various plans of constructing rail-roads, now in use, are particularly described in the body of the work. See Columbia Rail-road, Pennsylvania.

Rails, iron bars of various shapes, used upon rail-roads.

Ravine, the valley, or gulley through which a stream flows.

Reservoir, an enclosure of water, artificially made in order to collect and retain it for the use of canals, mills and other purposes.

Retaining wall, a wall used for the support and maintenance of a body of earth, when circumstances render it expedient to slope it gradually down. Retaining walls are sometimes used where land is valuable, and are battered on the outside face from 1 to $1\frac{1}{2}$ inches to the foot ; the greatest degree of batter (which is usually curved) being given to the foot of the wall. Counterforts are generally carried up at the back of the wall, and piers in the front of it.

Rise, ascent; in civil engineering, it means an upward progress from one level to another.

Roads, the most general means of communication from one place to another.

Road bed, that part of a rail-road upon which the superstructure reposes.

Roman cement, a water cement, generally used with an equal portion of sharp sand.

Rope, for inclined plane, see Description of Columbia Railroad, art. Pennsylvania.

Rubble work, a rough description of masonry, the stones being laid in as regular courses as found convenient, and well flushed with mortar; and occasional banders, running through the whole thickness of the wall, are inserted, to tie the whole together, (which are more needed in this kind of masonry than in any other); chain band may also be used in rubble work, with great advantage, if many openings are required to be left.

S.

Safety car, a machine which follows or precedes rail-road cars in their passage of inclined planes, and prevents their descent in case of accident to the machinery, or otherwise.

Sand, a granular mineral substance insoluble in water.

Sandstone or Freestone, a durable stone when of good quality. It is generally found stratified, and as such, is easily cut into any form ; each stratum varies in thickness from about that of a slate to many feet, being, at different places, siliceous, argillaceous and calcarious. It varies in its component parts.

Scarfing, the joining of two pieces of timber together by oblique cuts which are usually 3 or 4 times the width of the face of the beam.

GLOSSARY.

Scraper, a machine drawn by horses or oxen, for excavating trenches, for canals, rail-roads, &c.

Section, of a canal or rail-road is its profile as presented by a transverse cut.

Self-acting plane, effects the ascent of one carriage by the descent of another, more heavily laden; the impelling force being that of gravity.

Shaft, a vertical sinking or well, excavated for the purpose of working a tunnel, or for ascertaining the nature of the ground.

Shale, indurated or compact slaty clay.

Shank, that part of a spike between the head and point.

Sheet piling, a row of timbers driven into the earth side by side, which are sometimes grooved and tongued together, and used for protecting foundation walls from the effects of water.

Sheeves or friction rollers, small wheels made of cast iron and used on inclined planes for the purpose of receiving the rope.

Side cuts, are those lateral canals which diverge from any leading canal.

Side cutting, material for embankment taken from the side of the line.

Side lines, are short sections of a rail-road on either side of a main line, with which they communicate by " turn-outs." They enable carriages to pass each other.

Sidling, excavation along parallel slopes.

Silex, flint.

Silicious, flinty; a rock of which silex is the chief component.

Sills, blocks of wood or stone upon which the string pieces, &c. of a rail-road are placed.

Skew back, the course of masonry forming the abutments to a segmental arch or to the cast iron ribs employed in bridges.

Skew bridge, differs from the ordinary draw-bridge only in the action of the draw, which opens upon a pivot, with a horizontal motion.

Slackwater navigation, is effected by means of dams which back the water and form pools of the required height. These occur more or less frequently according to the inclination of the bed of the stream. The pools thus created are connected by means of lift locks and short canals, by which the boats pass from one to another.

Sleepers, upon rail-roads, are generally of wood, about 5 by 9 inches and 9 feet long; they are placed across the track, and three feet apart from centre to centre.

Slide, a part of a forcing pump.

Slides, those portions of the canal and rail-road banks which become detached and precipitate themselves upon the line.

Slips, connecting sections.

Slope, declivity.

Slough, a small drain at the top of an embankment, to pass the water into the side drain.

Sluice, a water-gate; a flood-gate; a vent for water.

Sofite, the underside of any over-hanging erection, as the intrados of an arch, the underside of a cornice, &c.

Span, is the horizontal distance from one end of an arch to the other or its chord.

Specific gravity. The specific gravity of a body is its weight in relation to the weight of another body; pure water is now the standard of comparison, a cubic foot of which being assumed as unity.

Spike, a large nail, with which the plate rail is usually secured to the sleepers in rail-roads.

Splicing plates, metallic plates applied to the joinings of timber to hold the parts together.

Spoil bank, surplus excavation which is laid by the side of a line of rail-road to save expense of removal.

Stone, an aggregation of several mineral substances. There are three classes of stones (though some partake of all), viz.— silicious, argillaceous and calcareous.

Stop planks, dams on the line of a canal to prevent the loss of water in case of accident.

Stationary engines, are used for effecting the ascent and descent of carriages along inclined planes.

Stationary plane, a plane worked by a stationary engine and rope.

Steam (force of), at a temperature of 212 degrees, the elastic force of steam is equal to a pressure of $14\frac{1}{2}$ pounds on a square inch, and supports a column of mercury of the same base, 30 inches high; at 250 degrees, it supports 58 inches of

GLOSSARY. 261

mercury; at 300 degrees, 112 inches; and at 325 degrees, 141 inches.

Steam engine, a machine originally contrived for raising water by means of the expansive force of steam, produced from water or other liquids in a state of ebulition. A good steam engine, with a cylinder of 6 inches diameter, is about equal to the force of one horse, and with a cylinder of 14 inches diameter, it is called a ten-horse power.

Steam whistle, a device for warning persons when the engine is approaching. It consists of a pipe, at the top of the boiler, with a cock which when turned, the steam escapes with a loud hissing noise.

Strata, successive layers spread one over another.

Stratum, singular of Strata, a bed; a layer.

String course, a term applied to a course of masonry or brickwork projecting from the face of the wall.

String pieces, wooden rails upon which the iron bars of rail-roads are placed.

Summit level, the highest level.

Superstructure of a rail-road, consists of sills, cross-ties, string-pieces, iron rails, &c.

Switch, an iron rod placed at the intersections of rail-tracks, for the purpose of guiding the wheels of carriages in their passage from one track to another.

Syphon, a curved or bent pipe, designed chiefly for the conveyance of water to a distance over intervening eminences. This simple but, at the same time, interesting hydraulic instrument is but little used, probably owing to the difficulty of exhausting the air, when it is of a large caliber.

T.

Trail, one of the numerous kinds of iron rails now employed upon rail-roads. It derives its name from its resemblance to that letter.

Teaming, the operation of leading the earth or excavation from a cutting to the embankment.

Tender, a wagon accompanying a locomotive engine for the conveyance of fuel, water, &c.

Terminus, the end, the extreme point.

Terminal plane, the plane at either end of a rail-way.

Terminii, the ends.

Terrace, a small mount of earth covered with grass; a gallery or balcony.

Tide locks, canal locks, which unite rivers with canals; they are generally employed in regulating the supply of water, and to prevent its encroachment upon the works.

Timber, a term applied to trees after they are felled.

Tonnage, burthen, the weight or measure of any species of merchandize.

Tow path, a narrow road, travelled by horses in dragging boats along a canal.

Track, a road, now generally applied to the superstructure of rail-roads.

Tractive power, the power of draught required to overcome the friction or resistance of a canal or rail-way.

Traction, the amount of tractive power necessary to overcome the resistance on a rail-road or canal.

Train, a regular succession of carriages, fastened together, and drawn by a locomotive engine or other motive power.

Tram-way. The modern tram-way is formed of plates of cast iron, 4½ inches wide, and 1 inch thick, and laid in 3 feet lengths; the plates have an upper vertical guide flange, 2 inches high, and a fish-bellied lower flange on the other side. The guide rails are 4 feet apart, and the space between each line is 5 feet; the plates are bedded on stone blocks, and fastened down by iron spikes driven into wooden plugs, which are let into the blocks vertically.

Tram-ways are sometimes constructed of stone or blocks of granite, 16 inches wide, 12 inches thick, and 5 or 6 feet long, the space between the trams being filled with paving. A sample of this kind of tram-way may be seen in Arch street in Philadelphia. Tram-ways, originally constructed of timber, were introduced into England about the year 1600.

Trench, a pit or ditch.

Trench walls, the masonry at the sides of trenches.

Trestles, the support of bridges, and other structures.

Trunk, the main or principal line of a canal or rail-road.

Tunnel, a subterraneous arch-way or gallery, excavated through a hill for the passage of a canal or rail-road.

GLOSSARY. 263

Turn-out, rails which diverge from the main line of a rail-road, and lead to the side lines.

Turn-plate or *Turn-table*, a platform which turns upon a pivot, for removing rail-road carriages from one track to another, they are generally used for crossing at right angles with each other.

Truck, a stage or platform on wheels, and used on rail-roads for the conveyance of ordinary carriages. In England it is customary for the mail coaches to be thus conveyed, the passengers and luggage remaining in their places.

V.

Valve, in hydraulics and pneumatics, is a kind of lid or cover to a tube, vessel or orifice, contrived to open only one way, and either admits the entrance of a fluid into the tube or vessel, and prevents its return, or allows its escape and prevents its re-entrance.

Variation of the compass, is the angle which a magnetic needle, suspended at liberty, makes with the meridian line on a horizontal plane; or an arch of the horizon, comprehended between the true and magnetic meridian. In nautical language it is commonly called north-westing or north-easting.

Velocity, speed, swiftness, quick motion; in its application to a rail-road train, it means the degree of speed with which it is propelled. The average speed of the English passenger cars is about 25, while that on the American rail-road does not exceed 16 miles an hour.

Ventilators, in covered aqueducts, are apertures by which the atmospheric air escapes when displaced by water.

Vertical, perpendicular to the horizon.

Vertex, zenith, a point, top of a hill.

Viaduct, a rail-road bridge.

W.

Water stations, places where locomotives obtain their supplies of water.

Water works, in general, denote all manner of machines moved by, or employed in raising or sustaining water.

Web, the outer projection of a rail, intended to prevent the wheels of carriages from running off the track.

Wedge, one of the five mechanical powers, or simple engines, being a geometrical wedge, or very acute triangular prism, applied to the splitting of wood, rocks, or raising great weights.

Waste weir, a water guage ; a cut at the side of a canal by which the surplus water of canals is carried off. The front of the cut next the canal is sometimes faced with masonry, which is carried from the bottom of the canal to the prescribed top water line, when the height of the water exceeds this, it passes off.

THE END.

INDEX.

A.

Adonirac R. R., N. Y. . Page	81
Advertisement . . .	9
Akron and Perrysburg R. R., O.	213
Akron and Defiance R. R., O.	213
Alabama, Florida, and Georgia R. R., Al. . .	180
Albany and Weststockbridge R. R., N. Y. . .	76
Albion and Tonawanda R.R., N. Y.	81
Atchafalaya R. R., La. .	189
Alexandria Canal, Dis. Col. .	159
Allegany Portage R. R. .	126
Allegan and Marshall R. R., Mic.	217
Alton & Mount Carmel R. R., Il.	196
Alton and Erie, Il. .	197
Alton and Paris R. R., Il. .	197
Alton and Springfield R. R., Il.	197
Amoskeag Canal, N. H. .	33
Andover and Wilmington R. R., Mass. . .	37
Andover & Haverhill R. R., Mass.	38
Annapolis & Elkridge R. R., Md.	157
Ann Arbor and Monroe R. R., Mich. . .	218
Arkansas, State of . .	214
Ashtabula and Liverpool R. R., O.	213
Athens Branch Canal, O. .	210
Athens Branch R. R., Ga. .	173
Attica and Buffalo R. R., N. Y.	80
Attica and Sheldon R. R., N. Y.	81
Auburn & Rochester R. R., N. Y.	79
Auburn Canal and R. R., N. Y.	81
Auburn and Lapeer R. R., Mich.	218
Augusta Branch R. R., Ga. .	176
Aurora and Buffalo R. R., N. Y.	81

B.

Bald Eagle and Spring Creek Navigation, Pa. .	112
Baltimore and Ohio R. R., Md.	149
Baltimore and Susquehanna R. R., Md. . .	155
Baltimore and Port Deposite R. R., Md. . .	151
Bangor and Orono R. R., Me.	29

Barataria Navigation, La. .	190
Bath R. R., N. Y. . .	81
Baton Rouge & Clinton R. R., La.	189
Bayou Sara R. R., La. .	189
Beaver Meadow R. R., Pa. .	143
Belmont and Dodgeville R. R., Wis. . .	220
Belmont & Dubuque R. R., Wis.	220
Belleville R. R., Il. .	197
Bellefontaine and Perrysburg R. R., O. . .	213
Belleville and Bolivar Canal, O.	211
Blackstone Canal, Mass. .	43
Black River Canal, N. Y. .	56
Black River R. R., N. Y. .	81
Bloomington and Peoria R. R., Il.	197
Boston and Lowell R. R., Mass.	35
Boston & Worcester R. R., Mass.	38
Boston & Providence R. R., Mass.	41
Bow Canal, N. H. . .	32
Brandon and Mobile, Miss. & Al.	184
Brattleboro and Bennington R. R., Vt. . .	34
Brewertown and Syracuse R. R., N. Y. . .	82
Bridgeport & Sandusky City, O.	213
Bridgeport and Sawpits R. R., Con.	48
Brooklyn Ft. and C. Island R. R., N. Y. . .	81
Brunswick and Florida R. R., Ga.	178
Brunswick Canal, Ga. .	178
Buck Mountain R. R., Pa. .	112
Buffalo and Batavia R. R., N. Y.	82
Buffalo and Black Rock R. R., N. Y. . .	80
Buffalo and Erie R. R., N. Y.	82
Buffalo and Niagara Falls R. R., N. Y. . .	80

C.

Cahawba and Marion R. R., Al.	182
Calais and Milltown R. R., Me.	30
Camden and Amboy R. R., N. J.	84
Camden and Woodbury, R. R., N. J. . .	91
Cape Fear R. Navigation, N. C.	169

266 INDEX.

Carrollton and Lodi R. R., O.   213
Carondelet Canal, La. . .   190
Carbondale and Honesdale R. R.
  Pa.   143
Casadaga and Erie R. R., N. Y.   82
Catawba Navigation, N. C. & Va. 172
Catawissa and Towanda R. R., P. 140
Catawissa R. R. (See Little Schuylkill and Susquehanna R. R.)
Catskill and Canajoharie R. R.,
  N. Y.   76
Cayahuga Falls and Cleveland R.
  R., O.   213
Cayuga and Seneca Canal, N. Y.   56
Central R. R., Geo. . . .   174
Central R. R., Il. . . .   196
Central Canal, Ind. . .   198
Central Canal, Pa. . .   97
Central R. R., Vt. . . .   34
Central R. R., Mich. . .   215
Chagrine and Holmes Canal, O.   211
Chambersburg and Pittsburg R.
  R., Pa.   132
Champlain Canal, N. Y. . .   54
Charleston and Elyria R. R., O.   213
Charleston and Ashland R. R., O. 213
Charlestown Branch R. R., Mass. 38
Chattahoochee R. R., Geo.   178
Chemung Canal, N. Y. . .   57
Chemung and Ithaca R. R., N. Y. 82
Chenango Canal, N. Y. . .   55
Cherry Valley and Susquehanna
  R. R., N. Y. . . .   82
Cherry Stone and Maryland line
  R. R. Va. . . .   167
Chesapeake and Delaware Canal,
  Del.   148
Chesapeake and Ohio Canal, Md. 158
Chesterfield R. R., Va. .   165
Chicago and Des Plaines R. R.,
  Il.   197
Chillicothe and Cincinnati R. R.,
  O.   213
Cincinnati and Harrison Canal,
  O.   211
Cincinnati and Indianapolis R.
  R., O. & Ind. . . .   213
Circleville and Cincinnati, R. R.,
  O.   213
City Point R. R., Va. .   164
Cleveland and Cincinnati R. R.,
  O.   213
Cleveland and Pennsylvania line
  R. R., O. . . .   213
Cleveland and Warren R. R., O. 213
Cleveland and Franklin R. R., O. 213
Clinton and Chippeway Canal, O. 211
Clinton and Kalamazoo R. R.,
  Mich.   219

Clinton and Lower Sandusky R.
  R., O.   213
Clinton and Adrian R. R., Mich. 219
Clubfoot Canal, N. C. . .   169
Coeymans R. R., N. Y. . .   82
Codorus Navigation, Pa. .   112
Colbert's Shoals Canal, Al. .   183
Cold Springs R. R., N. Y.   82
Columbia and Philadelphia R. R.
  Pa.   113
Columbus and Chattahoochee R.
  R., Geo.   178
Columbus and Pensacola R. R.,
  Fl.   179
Columbus and Aberdeen R. R.,
  Miss.   184
Columbus and Lower Sandusky
  R. R., O.   213
Columbus and Springfield R. R.,
  O.   213
Columbus Branch Canal, O. .   210
Columbus and Delaware C., O.   211
Columbus and Big Spring R. R.,
  O.   213
Cooperstown and Cherry Valley
  R. R., N. Y. . . .   82
Connecticut, State of . .   47
Connecticut and Passumpsic R.
  R., Vt.   34
Conneaut and Penn. R. R., O.   213
Conestoga Canal, Pa. . .   112
Conewango Canal, N. Y. .   57
Condensed summary of the R.
  R. and Canals of U. S. .   223
Constantine and Niles R. R., Mich. 218
Corning and Blossburg R. R., Pa. 139
Covington and Latona R. R., Ky. 193
Coxsackie and Schenectady R. R.
  N. Y.   82
Croton Aqueduct, N. Y. .   59
Crooked Lake Canal, N. Y. .   57
Cumberland and Oxford Canal,
  Me.   29
Cumberland Valley R. R., Pa.   131

D.

Dansville and Rochester R. R.,
  N. Y.   82
Dansville Branch Canal, N. Y.   59
Danville and Wythe R. R., Va.   166
Danville and Pottsville R. R., Pa. 137
Delaware, state of . .   147
Dedham Branch R. R., Mass.   42
Delaware and Columbia Canal, O. 211
Delaware & Hudson Canal, N. Y. 58
Delaware R. R., N. Y. .   82
Delaware R. R., Del. . .   148
Delaware and Raritan Canal, N. J. 84
Detroit and Pontiac R. R., Mich. 217

INDEX. 267

Detroit and Owasso R. R., Mich. 218
Detroit and Utica R. R., Mich. 218
Detroit and Monroe R. R., Mich. 218
Detroit & Maumee R. R., Mich. 218
Detroit and Shiawassee R. R., Mich. 218
Dismal Swamp Canal, Va. . 161
Drehr's Canal, S. C. . . 172
Dresden Branch Canal, O. . 210
Dutchess R. R., N. Y. . . 82

E.
Eastern Rail-road, Me. . 32
Eastern R. R., Mass. . . 35
Eastern Shore R. R., Md. . 157
East Florida R. R., Fl. . . 179
Eastport Branch Canal, Me. 210
Edwardsville and Shawneetown R. R., Il. . . . 196
Elizabethport and Somerville R. R., N. J. . . . 94
Enfield Canal, Con. . . 49
Erie Canal, N. Y. . . 52
" " Enlargement . 53
Erie and Cattaraugus R. R., N. Y. 82
Erie and Kalamazoo R. R., Mich. 217
Erie and Ohio R. R., O. . 213

F.
Fairfield County, R. R., Con. . 48
Fairmount Water Works, Pa. 104
Falmouth & Lexingt'n R. R., Ky. 193
Farmington Canal, Con. . 45
Fishhouse and Amsterdam R. R., N. Y. . . . 82
Florida, Territory of . 179
Franklin and Wilmington R. R., O. . . . 213
Franklin R. R., Pa. . . 131
Fredericksburg and Charlottesville R. R., Va. . . 166
Fredonia and Van Buren Harb. R. R., N. Y. . . . 82

G.
Galena and Chicago, Il. . 197
Georgia, State of . . 173
Georgia R. R., Ga. . . 173
General View . . . 11
Genesee & Pittsford R. R., N. Y. 82
Genesee and Cattaraugus R. R., N. Y. 82
Genesee Valley Canal, N. Y. 59
Geneva and Canandaigua R. R., N. Y. 82
Gettysburg R. R., Pa. . 145
Gibraltar & Clinton R. R., Mich. 218
Gilboa R. R., N. Y. . 82
Glossary . . . 235

Goshen and N. Jersey R. R., N. Y. 82
Granville Branch Canal, O. . 210
Grand Gulf and Port Gibson R. R., Miss. . . . 184
Grand R. and Saginaw Canal, Mich. 219
Great Western R. R., Mass. 39
Great Au Sable R. R., N. Y. . 82
Green River Navigation, Ky. 192
Greene R. R., N. Y. . . 82
Greensville R. R., Va. . 164

H.
Hampshire and Hampden Canal, Mass. 43
Harrisburg and Lancaster R. R., Pa. 130
Harlem Canal, N. Y. . 59
Harlem R. R., N. Y. . . 72
Hartford and Springfield R. R., Con. 48
Hazelton R. R., Pa. . . 143
Havre Branch R. R., Mich. . 216
Henderson and Nashville R. R, Ky. 193
Herkimer & Trenton R. R., N. J. 82
Highwassee R. R., Ten. . 191
Hookset Canal, N. H. . 33
Homer and Union Canal, Mich. 219
Honeyoye R. R., N. Y. . 82
Housatonic R. R., Con. . 48
Hopkinsville and Cumberland R. R., Ky. . . . 193
Hudson & Berkshire R. R., N. Y. 75
Hudson & Delaware R. R., N. Y. 82
Huntsville Canal, Al. . 183
Huron Navigation, Mich. . 219

I.
Illinois, State of . . 194
Illinois and Michigan Canal, Il. 195
Illinois and Mine Bluff R. R., Il. 197
Indiana, State of . 198
Iowa, Territory of . . 220
Ithaca and Auburn R. R., N. Y. 82
Ithaca and Geneva R. R., N. Y. 82
Ithaca and Owego R. R., N. Y. 80
Ithaca & Pt. Renwick R. R., N. Y. 82

J.
Jackson & Brandon R. R., Miss. 184
Jacksonville and Tallahassee R. R., Fl. . . . 179
Jacksonville & Augusta R. R., Il. 197
James River & Kanawha Canal and R. R., Va. . . 160
Jamesville R. R., N. Y. . 82
Jobstown Branch R. R., N. Y. 90
Johnstown R. R., N. Y. . 82

Jordan and Skaneateles, N. Y.		82
Jeffersonville and Indianapolis, Ind.	. . .	201

K.

Kalamazoo and Dexter Canal, Mich.	. . .	219
Kalamazoo and Michigan R. R., Mich.	. . .	219
Kalamazoo and S. Black River R. R. Mich.	. .	218
Kanawha and Coal river R. R. Va.		167
Kentucky, State of	. .	192
Kentucky Navigation, Ky.	.	192
Kingston R. R., N. Y.	. .	83

L.

Lackawaxen Canal, Pa.,	. .	112
La Chine Canal, Can.	. .	222
La Grange and Memphis R. R., Ten.	. . . .	191
Lafayette and Michigan R. R., Ind.	. . . .	201
Lafayette and Danville R. R., Ind.	. . .	201
Lafontaine and Winnebago R. R., Wis.	. .	220
Lake Borgne R. R., La.	.	189
Lake Drummond Canal, N. C.		168
Lake Veret Canal, La.	. .	190
Lake Wimico and St. Joseph R. R.,	. . . .	179
Lancaster Branch Canal, O.	.	210
La Prairie and St. Johns R. R., Can.	. . . .	222
Lawrenceville and Indianapolis R. R., Ind.	. .	201
Lehigh Navigation, Pa.	. .	110
Lehigh and Susquehanna R. R., Pa.	. . . .	143
Lewistown R. R., N. Y.	.	82
Lexington and Ohio R. R., Ky.		193
Licking Navigation, Ky.	.	192
Lima and Shanesville R. R., O.	.	213
Little Miami R. R., O.	.	212
Little Schuylkill R. R., Pa.	.	137
Little Schuylkill and Susquehanna R. R., Pa.	. . .	138
Lockport and Batavia R. R., N. Y.	. . .	82
Lockhart Canal, S. C.	.	172
Lockport and Niagara Falls R. R., N. Y.	. . .	80
Lockport and Youngstown R. R., N. Y.	. . .	82
Long Island R. R., N. Y.	.	70
Loricks Canal, S. C.	.	172
Louisa R. R., Va.	.	165
Louisiana, State of,	.	185

Louisiana and Columbia R. R., Mo.	. . .	214
Louisville, Cincinnati and Charleston R. R., N. C., S. C., Ten., Ky.		171
Louisville and Bushville R. R., Ky.		193
Louisville and Knoxville R. R., Ky.	. . .	193
Louisville and Portland Canal, Ky.		193
Lower Sandusky and Tyemochte Canal, O.	. . .	211
Lykens valley, R. R. Pa.	. .	143
Lynchburg and Tennessee R. R., Va.	. . . .	166
Lynville and Jacksonville R. R., Ill.	. . . .	198

M.

Macon and Talbottom R. R., Ga.		177
Mad River and Sandusky City R. R., O.	. . .	212
Madison and Indianapolis R. R., Ind.	. . . .	201
Madisonville and Pond River R. R., Ky.	. . .	193
Mahoning Canal, O. & Pa.		211
Maine, State of	. .	29
Malden R. R., N. Y.	.	82
Manchester and Benton R. R., Miss.	. . .	184
Manheim and Salisbury R. R., N. Y.	. . .	83
Mansfield and New Haven R. R., Con.	. . .	213
Marietta R. R., Pa.	.	133
Maryland, State of	.	149
Maryland Canal, Md.	. .	159
Massillon and Ohio R. R., O.		213
Massachusetts, State of	.	35
Matanzas and Halifax Canal, Fl.		179
Mauch Chunk R. R., Pa.	.	142
Mayville and Portland R. R., N. Y.	. . .	83
Medina and Canandaigua R. R., Mich.	. . .	218
Medina and Darien R. R., N. Y.		83
Medina and Lake Ontario R. R., N. Y.	. . .	83
Melmore and Republic R. R., O.		213
Miami Canal, O.	. .	210
Michigan, State of	.	215
Middlesex Canal, Mass.	.	42
Milan Canal, O.	.	211
Milan and Newark R. R., O.	.	213
Milan and Lebanon R. R., O.		213
Millbury Branch R. R., Mass.		39
Mill Creek R. R., Pa.	.	142
Millwaukee and Black R. Canal, Wis.	. . .	220
Millwaukee & Missis. R. R., Wis.		220

INDEX. 269

Mine Brook R. R., N. J. . 94
Mine Hill and Schuylkill Haven
  R. R., Pa. . . . 142
Mississippi, State of . 184
Mississippi R. R., Miss. . 184
Missouri, State of . . 214
Mobile & Cedar Point R. R., Al. 182
Mohawk & Hudson R. R., N. Y. 77
Monroe R. R., Ga. . . 177
Monroe and Cen. R. R., Mich. 218
Montague Canal, Mass. . 43
Montgomery and West Point R.
  R., Al. . . . 182
Morris Canal, N. J. . 84
Morris and Essex R. R., N. J. 93
Mottville and White Pigeon R.
  R., Mich. . . . 218
Mount Clemens and Saginaw R.
  R., Mich. . . . 218
Mt. Vernon and Mohiccon River
  Canal, O. . . . 211
Mount Carbon R. R., Pa. . 142
Muscle Shoals Canal, Al. . 183

N.

Naples and Jacksonville R. R., Il. 197
Nashua and Lowell R. R., N. H.
  and Mass. . . . 32
Nashville and Knoxville R. R.,
  Ten. 191
Natchez and Woodville R. R.,
  Miss. 184
Nesquehoning R. R., Pa. . 143
Newark R. R., N. Y. . . 83
Newark and Mt. Vernon R. R., O. 213
New Albany and Columbus R.
  R., Ind. . . . 201
New Bedford and Fall River R.
  R., Mass. . . . 42
New Castle and Frenchtown R.
  R., Del. . . . 147
New Castle and Wilmington R.
  R., Del. . . . 147
New Hampshire, State of . 32
N. Haven and Hartford R. R., Con. 47
New Haven and Monroeville R.
  R., O. . . . 213
New Jersey, State of . 84
New Jersey R. R., N. J. . 91
New Orleans and Nashville R. R.,
  La. Miss. and Ten. . 188
New York, State of . 50
New York and Albany R. R.,
  N. Y. 73
New York and Erie R. R., N. Y. 74
North Carolina, State of . 167
Northern Liberties and Penn T.
  R. R., Pa. . . . 130
Northern R. R., Mich. . 216

North West Canal, N. C. . 168
Norwalk and Huron R. R., O. 213
Norwich and Hartford R. R., Vt. 34
Norwich & Worcester R. R., Con. 47

O.

Ogdensburg and Champlain R.
  R., N. Y. . . . 81
Ohio, State of . . 202
Ohio and Erie Canal, O. . 209
Ohio R. R., O. . . 212
Orleans Bank Canal, La. . 188
Oswego Canal, N. Y. . 56
Oswego & Syracuse R. R., N. Y. 83
Oswego and Utica R. R., N. Y. 81
Otsego R. R., N. Y. . . 83
Owego and Cortland R. R., N. Y. 83

P.

Palmyra & Jacksonburg R. R.,
  Mich. 218
Paterson and Hudson R. R., N. J. 90
Pawtucket Canal, Mass. . . 42
Pekin R. R., Il. . . 197
Penfield R. R., N. Y. . 83
Pensacola and Mobile Branch R.
  R., Fl. 179
Pennsylvania, State of . 95
Pennsylvania & Indiana R. R., O. 213
Penn. Canal, Central Division 97
  do. do. Western do. . 98
  do. do. Susquehanna do. 99
  do. do. West Branch do. 99
  do. do. North   do.   99
  do. do. N. Br. Extension 99
  do. do. Delaware Division 100
  do. do. Beaver do.   100
  do. do. Conneaut Line . 102
  do. do. Erie Extension 101
  do. do. Franklin Line . 101
  do. do. French Cr. Feeder 101
  do. do. Lackawana do. 101
  do. do. Wisconisco Exten. 101
  do. do. Sinnemahoning Ex. 99
  do. do. Tangascootac Ex. 99
  do. do. Lewisburg Side Cut 99
  do. do. Bald Eagle " " 99
  do. do. Allegany Branch 98
  do. do. Johnstown Feeder 98
  do. do. Roystown Feeder 98
Peoria and Warsaw R. R., Il. 197
Petersburg and Roanoke R. R.,
  Va. 164
Petersburg and Farmville R. R.,
  Va. 166
Philadelphia, Germantown and
  Norristown R. R., Pa. . 144
Philad. and Reading R. R., Pa. 133
Philadelphia Rail-roads, Pa. 129

INDEX.

Philad. and Trenton R. R., Pa. 144
Philad. & Wilmington R. R., Pa. 144
Pine Grove R. R., Pa. . 144
Pittsburg Extension of Sunbury
  and Erie R. R., Pa. . 141
Pittsfield and Weststockbridge
  R. R., Mass. . . 42
Pontchartrain R. R., La. . 188
Portage R. R., Ky. . . 193
Portage R. R., Pa. . . 126
Port Hudson, Jackson and Clinton R. R., La. . . 189
Portland and Augusta R. R. Me. 31
Portland and Bangor R. R., Me. 30
Portland and Dover R. R., Me. 31
Portland, Saco, and Portsmouth
  R. R., Me. . . . 30
Portland and Quebec R. R., Me. 30
Portsmouth and Roanoke R. R.,
  Va. 165
Port Kent and Keesville R. R.,
  N. Y. . . . 81
Princeton and Deer Creek R. R.,
  Miss. 184
Providence and Stonington R. R.,
  R. I. 44

Q.

Quebec and St. Johns R. R. Can. 222
Quincy R. R. Mass. . . 42
Quincy and Danville R. R., Ill. 197

R.

Raleigh and Columbia R. R., N. C. 168
Raleigh and Gaston R. R., N. C. 168
Renssclaer and Saratoga R. R.,
  N. Y. . . . 76
Rhode Island, State of . . 44
Richmond and Danville R. R., Va. 166
Richmond and Petersburg R. R.,
  Va. 163
Richmond, Fredericksburg and
  Potomac R. R., Va. . 162
Richmond and Yorktown R. R.,
  Va. 167
Richmond and Miami R. R., O. 213
Rideau Canal, Can. . 221
Rivanna R. R., Va. . 166
River Raisin and Grand Riv. R.
  R., Mich. . . . 218
River Raisin and L. Erie R. R.,
  Mich. . . . 218
Roanoke Navigation, Va. . . 169
Rochester and Charlotte R. R.,
  N. Y. . . . 83
Rochester R. R., N. Y. . 80
Rochester and Lockport R. R.,
  N. Y. . . . 83

Rome and Port Ontario R. R., N.
  Y. 83
Romeo and Mt. Clemens R. R.,
  Mich. . . . 218
Room Run R. R., Pa. . . 142
Rutland and Connecticut R. R.,
  Vt. 34
Rutland and Whitehall, R. R. Vt. 83
Russelville and Clarksville R. R.,
  Ky. 193

S.

Saginaw R. R., Mich. . 218
Saginaw and Leroy R. R., Mich. 219
Salem Canal, N. J. . . 84
Saluda Canal, S. C. . . 172
Sandusky City and Monroeville
  R. R., O. . . . 212
Sandusky and Maumee R. R. O. 213
Sandy and Beaver Canal O. . 211
Santee Canal, S. C. . . 171
Saratoga and Fort Edward R. R.
  N. Y. . . . 83
Saratoga and Montgomery R. R.,
  N. Y. . . . 83
Saratoga and Schenectady R. R.,
  N. Y. . . . 78
Saratoga and Schuylerville R. R.,
  N. Y. . . . 83
Saratoga and Washington R. R.,
  N. Y. . . . 83
Savannah, Ogeechee and Alatamaha Canal, Geo. . . 178
Schoharie and Otsego R. R., N.
  Y. 83
Schuylkill Navigation, Pa. . 103
Schuylkill R. R., Pa. . . 142
Schuylkill Valley R. R., Pa. . 142
Scottsville and Onondaga R. R.,
  N. Y. . . . 83
Scottsville and Leroy R. R., N.
  Y. 83
Sebasticook and Moorehead Canal, Me. . . . 31
Selma and Cahawba R. R., Al. 182
Selma and Tennessee R. R., Al. 182
Sekonk R. R., Mass. . . 42
Sewals Falls Canals, N. H. . 33
Sharon and Root R. R., N, Y. . 83
Shelby and Belle River, R. R.,
  Mich. 218
Shelby and Detroit R. R., Mich. 218
Shiawassee Navigation, Mich. . 219
Skaneatcles R. R., N. Y. . 83
Smithfield and Winchester R. R.,
  Va. 167
Somerville Branch R. R., Ten. 191
South Carolina, State of . . 169
South Carolina R. R., S. C. . 169

INDEX. 271

South Hadley Canal, Mass. . 44
Southwark R. R., Pa. . . 130
Southern R. R., Mich. . . 215
Springfield and Liberty R. R., La. 189
Springfield and Carrollton R. R., Ill. 198
St. Andrews and Chipola Canal, Fl. 179
St. Augustine and Picolata R.R., Fl. 179
Staunton and Potomac R. R., Va. 166
Staunton and Scottsville R. R., Va. 166
State line and Ohio and Ashtabula R. R., O. . . 213
Staten Island R. R., N. Y. . 83
St. Clair and Romeo R. R., Mich. 217
St. Joseph and Tallahassee R. R., Fl. 179
Stillwater and Maumee R. R., O. 213
St. Louis and Iron Mountain R. R., Mo. 214
St. Louis and Jefferson R. R., Mo. 214
St. Louis and St. Charles R. R., Mo. 214
Strasburg R. R., Pa. . . 132
Suffolk Branch R. R. Va. . 166
Sugarloaf R. R., Pa. . . 112
Summary of Canals, &c., of the U. S. 223
Sunbury and Erie R. R., Pa. . 140
Susquehanna Canal, Pa. & Md. 113
Syracuse and Auburn R. R., N. 78
Syracuse and Cortland R. R., N. Y. 83
Syracuse and Onondaga R. R., N. Y. 83
Syracuse and Utica R. R., N. Y. 78
Syracuse and Stone R. R., N. Y. 83

T.

Tallahassee and St. Marks, R. R., Fl. 179
Tar River Navigation, N. C. . 169
Taunton Branch R. R., Mass. . 42
Taylorsville and Orange R. R.,Va. 161
Tennessee, State of . . 191
Terre Haute and Eel River Canal, Ind. 201
Terre Haute and Evansville R. R., Ind. 201
Toledo and Sandusky city R. R., O. 213
Tonawanda R. R., N. Y. . 79
Trenton Branch R. R., N. J. . 90
Trenton and New Brunswick R. R., N J. 93
Trenton and Sacketts Harb. R. R., N. Y. 83

Troy R. R., N. Y. . . 83
Troy and West Stockbridge R. R., N. Y. . . . 77
Tuscumbia, Courtland and Decatur R. R., Al. . . 182
Tyrone and Geneva R. R., N. Y. 83

U.

Ulster County R. R., N. Y. . 83
Unadilla and Schoharie R. R., N. Y. 83
Union Canal, N. H. . . 33
Union Canal, Pa. . . 109
Urbana and Columbus R. R., O. 213
Utica and Oswego Canal, N. Y. 57
Utica and Schenectady R. R., N. Y. 78
Utica and Susquehanna R. R. N. Y. 83

V.

Valley R. R., Pa. . . 130
Venice and Belleview R. R., O. 213
Vermillion and Birmingham R. R., O. 213
Vermillion and Ashland R. R.,O. 213
Vermont, State of . . 33
Vermont Central R. R. . . 34
Vicksburg and Clinton R. R., Miss. 184
Virginia, State of . . 160

W.

Wabash and Erie Canal, In. & O. 198
Walhonding Br., Canal, O. . 210
Warsaw and Leroy R. R., N. Y. 83
Warren Canal, O. . . 210
Warren County R. R., N. Y. . 83
Warrenton and Falmouth R. R., Va. 167
Warwick R. R., N. Y. . . 83
Washington Branch R. R., Md. 155
Washington Branch Canal, Md. 159
Watertown and Rome R. R., N. Y. 83
Watervliet and Schenectady R. R., N. Y. . . . 83
Wayne and Michigan City Canal, Ind. 201
Wayne and Piqua R. R., O. . 213
Waverly and Grand Prairie R. R., Il. 198
Weldon and Danville R. R., Va. 166
Weldon Canal, N. C. . . 168
Welland Canal, Can. . . 221
Wellsville and Fairport R. R., O. 213
Wellsville and Steubenville R. R., O. 213
Wetumpka R. R., Al. . . 128

INDEX.

West Chester R. R., Pa. . 130
Western and Atlantic R. R., Ga. 173
West Feliciana R. R., Miss. . 184
West Philadelphia Canal, Pa. 113
West Philadelphia R. R., Pa. 130
West Troy and Schenectady R. R., N. Y. 77
Whitehall and Saratoga R. R., N. Y. 77
White River Canal, Vt. . 33
White Water Canal, Ind. . 197
Whippany R. R., N. Y. . 94
Williamsport and Elmira R. R., Pa. 139
Wilmington and Raleigh R. R. Md. 167
Wilmington and Susquehanna R. R., Md. . . . 157
Winchester and Potomac R. R., Va. 165
Winyaw Canal, S. C. . . 172
Wisconsin, Territory of . . 220
Worcester and Hartford R. R., Con. 48

Y.

Yadkin Navigation, N. C., . 169
York and Wrightsville R. R., Pa. 132
Ypsilanti and Tecumseh R. R., Mich. 217
Ypsilanti and Raisin R. R., Mich. 218

Z.

Zanesville Branch Canal, O. . 210
Zanesville & Ohio River R. R., O. 213

Library of
Early American Business And Industry

I. John Leander Bishop, A HISTORY OF AMERICAN MANUFACTURES FROM 1608 TO 1860, with an introduction by Louis M. Hacker, 3 volumes.

II. Albert S. Bolles, THE INDUSTRIAL HISTORY OF THE UNITED STATES, Copious Illustrations, with an introduction by Louis M. Hacker.

III. Freeman Hunt, LIVES OF AMERICAN MERCHANTS, with an introduction by Louis M. Hacker, 2 volumes.

IV. George S. White, MEMOIR OF SAMUEL SLATER, Illustrated with engraving, woodcuts and folding diagram.

V. Rolla M. Tryon, HOUSEHOLD MANUFACTURES IN THE UNITED STATES, 1640-1860. A study in Industrial History.

VI. J. D. B. DeBow, THE INDUSTRIAL RESOURCES, etc. of the Southern and Western States, 3 volumes.

VII. TENCH COXE, A VIEW OF THE UNITED STATES OF AMERICA, with folding tables.

VIII. Charles F. Adams, Jr., and Henry Adams, CHAPTERS OF ERIE and other Essays.

IX. Stuart Daggett, RAILROAD REORGANIZATION.

X. Stuart Daggett, HISTORY OF THE SOUTHERN PACIFIC.

XI. Nelson Trottman, HISTORY OF THE UNION PACIFIC, a financial and economic survey.

XII. Howard D. Dozier, A HISTORY OF THE ATLANTIC COAST LINE RAILROAD.

XIII. Timothy Pitkin, A STATISTICAL VIEW OF THE COMMERCE OF THE UNITED STATES OF AMERICA.

XIV. Katherine Coman, ECONOMIC BEGINNINGS OF THE FAR WEST, 2 volumes.

XV. William R. Bagnall, THE TEXTILE INDUSTRIES OF THE UNITED STATES.

XVI. Witt Bowden, THE INDUSTRIAL HISTORY OF THE UNITED STATES.

XVII. Melvin T. Copeland, THE COTTON MANUFACTURING INDUSTRY OF THE UNITED STATES.

XVIII. Blanche E. Hazard, THE ORGANIZATION OF THE BOOT AND SHOE INDUSTRY IN MASSACHUSETTS BEFORE 1875.

XIX. Albert Gallatin, REPORT OF THE SECRETARY OF THE TREASURY ON THE SUBJECT OF ROADS AND CANALS, 1807.

XX. Henry S. Tanner, A DESCRIPTION OF THE CANALS AND RAILROADS OF THE UNITED STATES.

XXI. J. Warren Stehman, THE FINANCIAL HISTORY OF THE AMERICAN TELEPHONE AND TELEGRAPH COMPANY.

XXII. Kathleen Bruce, VIRGINIA IRON MANUFACTURE IN THE SLAVE ERA.

XXIII. Abraham Gesner, A PRACTICAL TREATISE ON COAL, PETROLEUM AND OTHER DISTILLED OILS, revised and enlarged by George W. Gesner.

XXIV. Alexander Hamilton, INDUSTRIAL AND COMMERCIAL CORRESPONDENCE OF ALEXANDER HAMILTON ANTICIPATING HIS REPORT ON MANUFACTURES, edited by Arthur H. Cole. With a Preface by Prof. Edwin F. Gay.

XXV. Lewis Henry Haney, A CONGRESSIONAL HISTORY OF RAILWAYS IN THE UNITED STATES, 2 volumes in one.

XXVI. Adam Seybert, STATISTICAL ANNALS. Quarto.

XXVII. Samuel Batchelder, INTRODUCTION AND EARLY PROGRESS OF THE COTTON MANUFACTURE IN THE UNITED STATES.

XXVIII. Tench Coxe, A STATEMENT OF THE ARTS AND MANUFACTURES OF THE UNITED STATES OF AMERICA FOR THE YEAR 1810.

XXIX. (Louis McLane), DOCUMENTS RELATIVE TO THE MANUFACTURES IN THE UNITED STATES (Executive Document No. 308, 1st Session, 22nd Congress.

XXX. B. F. French, THE HISTORY OF THE RISE AND PROGRESS OF THE IRON TRADE OF THE UNITED STATES.

XXXI. Frederick L. Hoffman, HISTORY OF THE PRUDENTIAL INSURANCE COMPANY OF AMERICA, 1875-1900.

XXXII. Charles B. Kuhlman, DEVELOPMENT OF THE FLOUR MILLING INDUSTRY IN THE UNITED STATES.

XXXIII. James Montgomery, A PRACTICAL DETAIL OF THE COTTON MANUFACTURE OF THE UNITED STATES OF AMERICA.

XXXIV. Henry Varnum Poor, HISTORY OF THE RAILROADS AND CANALS OF THE UNITED STATES.

XXXV. Henry Kirke White, HISTORY OF THE UNION PACIFIC RAILWAY.

XXXVI. Frank B. Copley, FREDERICK W. TAYLOR, FATHER OF SCIENTIFIC MANAGEMENT, 2 volumes.

XXXVII. Edward Winslow Martin, HISTORY OF THE GRANGE MOVEMENT: or, The Farmer's War Against Monopolies.